Eduardo Kac and Avital Ronell met at Dis Voir, Paris, on July 23, 2007.


*Avital Ronell translated*
by Paul Buck & Catherine Petit


*Special Thanks*
To Suzanne Doppelt for her instrumental help with everything.

*and*
Ruth Kafensztok
Stephan May
Alban Barré
Solal de Aïa
Magali Guyon
Philip Ross
George Gessert
Annick Bureaud
Hugues Marchal

# LIFE EXTREME

## AN ILLUSTRATED GUIDE TO NEW LIFE

EDUARDO KAC & AVITAL RONELL

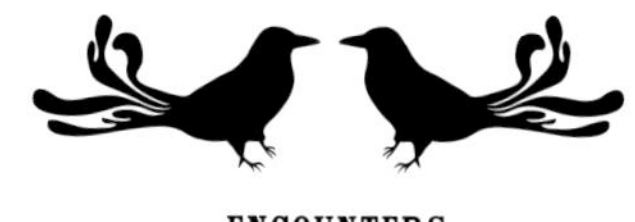

ENCOUNTERS

# LIFE EXTREME

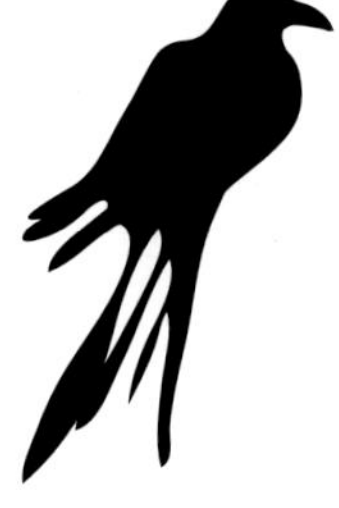

## ALPHABETICAL CHECKLIST

ALBA, THE BUNNY
AVITAL RONELL
ANTENNAPEDIA FLIES
BIZZARRIA ORANGE
BLUE ROSE
BUTTERFLIES WITH MODIFIED WING PATTERNS
CACTUS SCULPTURES
CATALINA MACAWS
CLONED DOG
COSMOPOLITAN CHICKEN
CUBIC WATERMELLON
DOLLY, THE CLONED SHEEP
EDUARDO KAC
FATHERLESS MOUSE
FEATHERLESS CHICKEN
FRIZZLED CHICKEN
GEEP
GLOFISH
HYBRID IRISES
HYPOALLERGENIC CAT
LAUFBERGER'S AXOLOTL
LIGER
LONGHORN CLONE
METHUSELAH WORM
MOUSE WITH HUMAN EAR
MUSHROOM SCULPTURES
NUDE MOUSE
QUAGGA
QUAIL-CHICKEN CHIMERA
RECONSTRUCTED AUROCHS
SAVANNAH CAT
SCHWARZENEGGER COW
SEMI-LIVING WORRY DOLLS
SPHYNX CAT
STEICHEN STRAIN DELPHINIUM
SYNTHETIC VIRUS
TRANSGENIC MONKEY
WHITE TIGERS
WOLPHIN
ZEEDONK

## CHRONOLOGICAL CHECKLIST

BIZZARRIA ORANGE (1674)
LIGER (1824)
FRIZZLED CHICKEN (1800s)
ZEEDONK (1800s)
LAUFBERGER'S AXOLOTL (1913)
RECONSTRUCTED AUROCHS (1920s)
CATALINA MACAW (1920s)
STEICHEN STRAIN DELPHINIUM (1936)
WHITE TIGERS (1950s)
AVITAL RONELL (1952)
EDUARDO KAC (1962)
SPHYNX CAT (1966)
NUDE MOUSE (1969)
QUAIL-CHICKEN CHIMERA (1975)
HYBRID IRISES (1979)
GEEP (1984)
WOLPHIN (1985)
SAVANNAH CAT (1986)
DOLLY, THE CLONED SHEEP (1997)
MOUSE WITH HUMAN EAR (1997)
SCHWARZENEGGER COW (1997)
COSMOPOLITAN CHICKEN (1998)
ANTENNAPEDIA FLIES (1998)
CACTUS SCULPTURES (1999)
ALBA, THE BUNNY (2000)
BUTTERFLIES WITH MODIFIED WING PATTERNS (2000)
SEMI-LIVING WORRY DOLLS (2000)
CUBIC WATERMELLON (2001)
TRANSGENIC MONKEY (2001)
MUSHROOM SCULPTURES (2001)
FEATHERLESS CHICKEN (2002)
SYNTHETIC VIRUS (2002)
LONGHORN CLONE (2002)
GLOFISH (2003)
METHUSELAH WORM (2003)
FATHERLESS MOUSE (2004)
BLUE ROSE (2004)
CLONED DOG (2005)
QUAGGA (2005)
HYPOALLERGENIC CAT (2006)

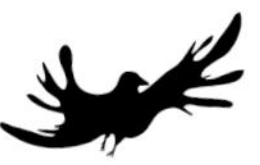

# BIOTECHNOLOGICAL CHECKLIST

**BACK-BREEDING**
QUAGGA
RECONSTRUCTED AUROCHS

**CHEMICAL SYNTHESIS**
SYNTHETIC VIRUS

**CHIMERISM**
GEEP
QUAIL-CHICKEN CHIMERA

**CLONING**
CLONED DOG
DOLLY, THE CLONED SHEEP
LONGHORN CLONE

**GRAFTING**
BIZZARRIA ORANGE
CACTUS SCULPTURES

**HORMONAL CHANGE**
LAUFBERGER'S AXOLOTL

**HYBRIDIZATION**
FEATHERLESS CHICKEN
WOLPHIN
HYBRID IRISES
SAVANNAH CAT
ZEEDONK
LIGER
CATALINA MACAWS
COSMOPOLITAN CHICKEN

**PHENOTYPICAL INTERVENTION**
CUBIC WATERMELLON
MOUSE WITH HUMAN EAR
BUTTERFLIES WITH MODIFIED WING PATTERNS
MUSHROOM SCULPTURE

**SELECTIVE BREEDING OF INDUCED MUTATION**
STEICHEN STRAIN DELPHINIUM

**SELECTIVE BREEDING OF SPONTANEOUS MUTATION**
FRIZZLED CHICKEN
WHITE TIGERS
SPHYNX CAT

NUDE MOUSE
SCHWARZENEGGER COW

**TISSUE CULTURE**
SEMI-LIVING WORRY DOLLS

**TRANSGENESIS / GENETIC PATHWAY**
ALBA, THE BUNNY
METHUSELAH WORM
HYPOALLERGENIC CAT
ANTENNAPEDIA FLIES
BLUE ROSE
TRANSGENIC MONKEY
GLOFISH
METHUSELAH WORM

**TRADITIONAL HUMAN REPRODUCTION**
AVITAL RONELL
EDUARDO KAC

# CONTEXTUAL CHECKLIST

**RESEARCH**
CLONED DOG
GEEP
QUAIL-CHICKEN CHIMERA
DOLLY, THE CLONED SHEEP
FATHERLESS MOUSE
SYNTHETIC VIRUS
NUDE MOUSE
LAUFBERGER'S AXOLOTL
METHUSELAH WORM
ANTENNAPEDIA FLIES
TRANSGENIC MONKEY
MOUSE WITH HUMAN EAR
METHUSELAH WORM

**ORNAMENTAL INDUSTRY**
FRIZZLED CHICKEN
WHITE TIGERS
BLUE ROSE
GLOFISH

**ENTERTAINMENT INDUSTRY**
LIGER
WOLPHIN
ZEEDONK

**DOMESTIC COMPANIONSHIP**
SPHYNX CAT
HYPOALLERGENIC CAT
CATALINA MACAWS
SAVANNAH CAT

**AGRICULTURE / BREEDING**
CUBIC WATERMELLON
FEATHERLESS CHICKEN
BIZZARRIA ORANGE
SCHWARZENEGGER COW
LONGHORN CLONE

**EXTINCT SPECIES RECOVERY**
QUAGGA
RECONSTRUCTED AUROCHS

**ART**
STEICHEN STRAIN DELPHINIUM
BUTTERFLIES WITH MODIFIED WING
PATTERNS
CACTUS SCULPTURES
HYBRID IRISES
MUSHROOM SCULPTURE
COSMOPOLITAN CHICKEN
SEMI-LIVING WORRY DOLLS
ALBA, THE BUNNY

**HUMAN FAMILY**
AVITAL RONELL
EDUARDO KAC

FLAVOR-SAVR TOMATO
FUNOTER
HEADLESS TADPOLES
HELA CELL
HUMAN-SHEEP CHIMERA
HYBRIDOMA
KHOFUTER
KHONORIK
KOHHOSIK
LEOPON
MALARIA-RESISTANT MOSQUITOES
MOUSE CHIMERA (INTERSPECIFIC)
MOUSE WITH HUMAN BRAIN CELLS
MOUSE THAT SWITCHES COLOR WITH
DIETARY SUPPLEMENT
MYCOPLASMA LABORATORIUM
PIGS WITH HUMAN BLOOD
PLANTS THAT GLOW WHEN TOUCHED
RAT-MOUSE CHIMERA
SERVICAL
SHEEP WITH HUMANIZED ORGANS
TIGON
TOAST OF BOTSWANA
TRANSGENIC BACTERIA AS BIOSENSORS
TRANSGENIC SALMON
ZONKEY
ZONY
ZORSE

# CHECKLIST OF BEINGS THAT ARE NOT IN THIS BOOK BOOK BUT COULD VERY WELL BE

ASTROPLANTS
BEEFALO
BLYNX
BRANDON BALLENGÉE'S BACKWARD-
BRED FROG
CAMA
CARAVAL
CC, THE CLONED CAT
CHAUSIE
DAVID POWELL'S MUTANT CATS
EURO CHAUS

IT CAN BE SAID THAT EVERYTHING POSSIBLE
DEMANDS EXISTENCE (...)
BUT IT DOES NOT FOLLOW FROM THIS
THAT ALL POSSIBLES EXIST;
THOUGH THIS WOULD FOLLOW
IF ALL POSSIBLE WERE COMPOSSIBLE.

- LEIBNIZ -

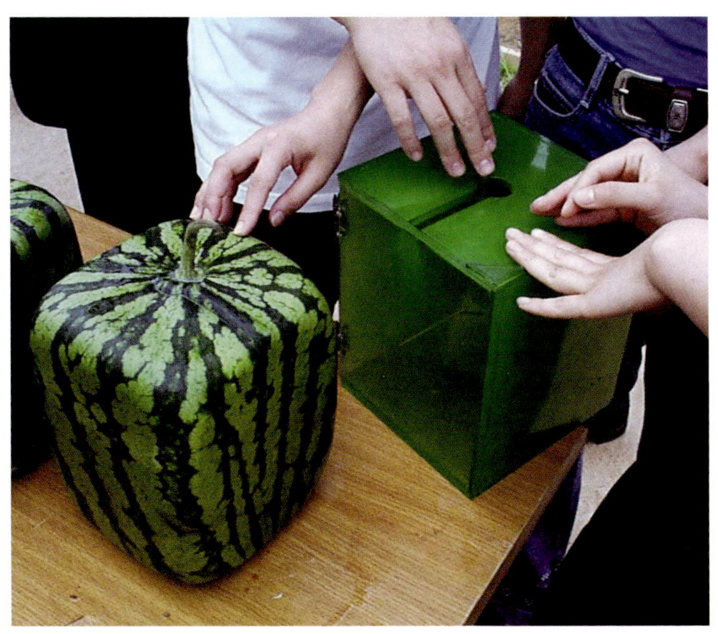

Faced by what concerns us, I would be tempted to start with a frank assertion of the impossible, allowing for all sorts of contaminations and hybridizations to prevail.
Both language and being are struck by the mutations before us. The enlivened images—which exceed the parameters of image by a long shot—put a dent in referential authority. Language itself balks and recedes, regressing into old habits and obsolesced paradigms. In a way, we are dealing with the drama of the referent where the positing power of language seems momentarily disabled: One thinks one sees a "cat" and one designates it as such, but words break down around this emergence. One persists in naming those referents as if one knew what was at hand. One persists in calling it a "cat," in saying it is a "mouse" that bears an over-proportioned ear. Language is returned to recognizable domains, its habitual comfort zones, without raising questions about the violence and lacerations done to it.

* * *

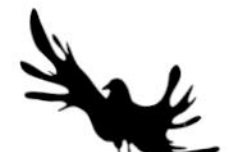

What is the living being? The stability of life or of the living is thrown off course—especially when one evokes technology or machinery, which lean more towards death. Looking at these images, these hyper-living forms, even though, born neither from Nature, nor from technology, what interests me beyond their strangeness, is the provocation from the past, the traumatic edges that appear. How are these beings installed in the psyche? What are the phobias that can be generated in so-called humans? When Heidegger wrote about "Dasein" instead of the human being, supplanting the one with the other, he shifted what one thought could be located as the human as no longer being anthropologically viable. The human started drifting away from the classical unity of the *anthropos*. Dasein calls upon a different calculus of being, no longer humanly centered.

* * *

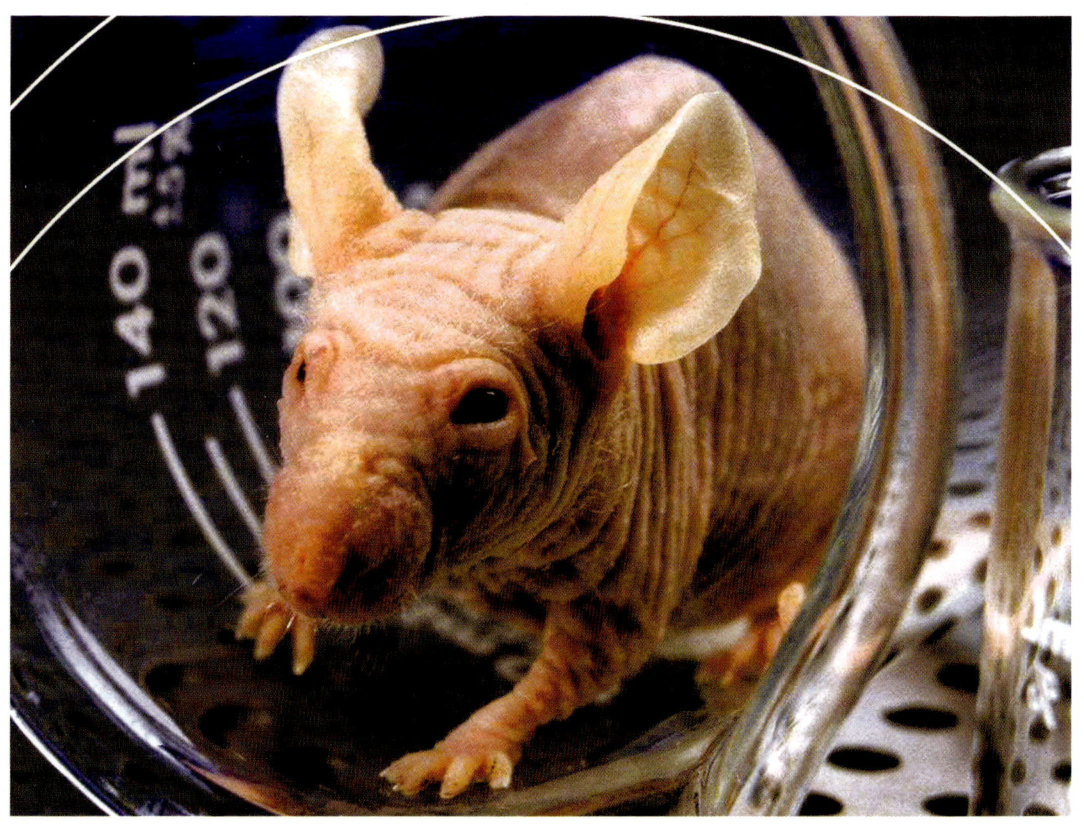

140 ml
120

I WILL NOW TELL YOU EVERY THING WHICH I MET
WITH ON THE MOON, THAT WAS NEW AND EXTRAORDINARY...
AMONGST THEM, WHEN A MAN GROWS OLD HE DOES NOT DIE,
BUT DISSOLVES INTO SMOKE, AND TURNS TO AIR. (...)

# HE WHO IS QUITE BALD, IS ESTEEMED A BEAUTY AMONGST THEM, FOR THEY ABOMINATE LONG HAIR;

WHEREAS, IN THE COMETS,
IT IS LOOKED UPON AS A PERFECTION;
AT LEAST, SO WE HEARD FROM SOME STRANGERS
WHO WERE SPEAKING OF THEM.

- LUCIAN OF SAMOSATA -

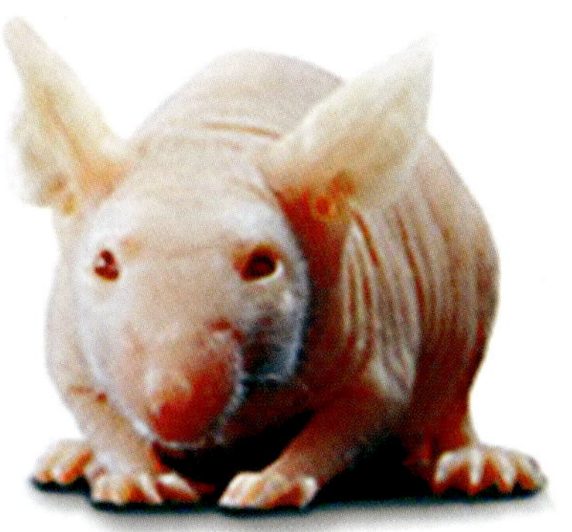

**PLATO DEFINED A HUMAN
AS A FEATHERLESS, BIPED ANIMAL
AND WAS APPLAUDED.**

**DIOGENES OF SINOPE PLUCKED A CHICKEN AND
BROUGHT IT INTO THE LECTURE HALL, SAYING:
"HERE IS PLATO'S HUMAN!"**

DIOGENES LAERTIUS -

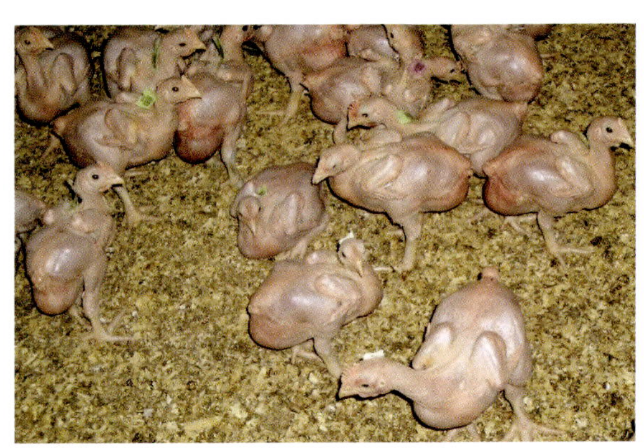

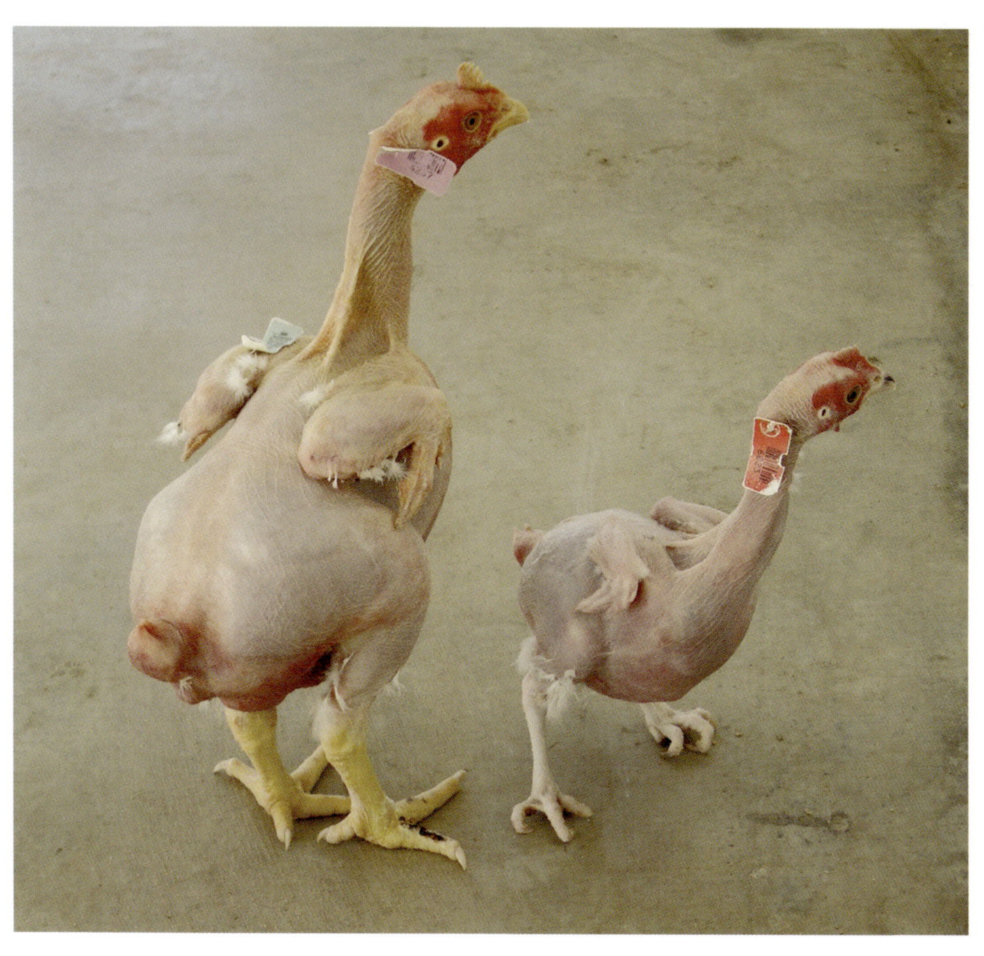

The "chicken," reduced to its thingness or as life exposed to its bareness, points to another history. It's impossible for me not to see in this chicken stripped completely naked-humiliated, cadaveric, walking around like a living dead, a kind of unconscious comment on the predicament of *human* destitution. That image makes me see something else, something other than itself, lesser and greater than itself. Of course, you could say that I project, that these are signs and affects, which I graft onto the stripped down thingness of a bald "chicken." In English, "chicken" underscores fearfulness. I project, let's say, a decidedly human quality. But that other thing, which is inexpressible and which passes through this image, is a violence, and not only because that featherless being is presented as "edible," ready for consumption, obscenely expedient, thoroughly industrialzied. What is shown there is the being that is caged, cornered, shamed, questioning.

* * *..

I allow myself to read a moment in my ownmost biography through the "featherless biped,"
my companion species. Maybe it is not a matter of *my* biography, after all. Even worse: I
recognize someone I know in the haggard exposition of shorn being.

* * *..

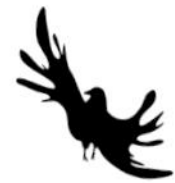

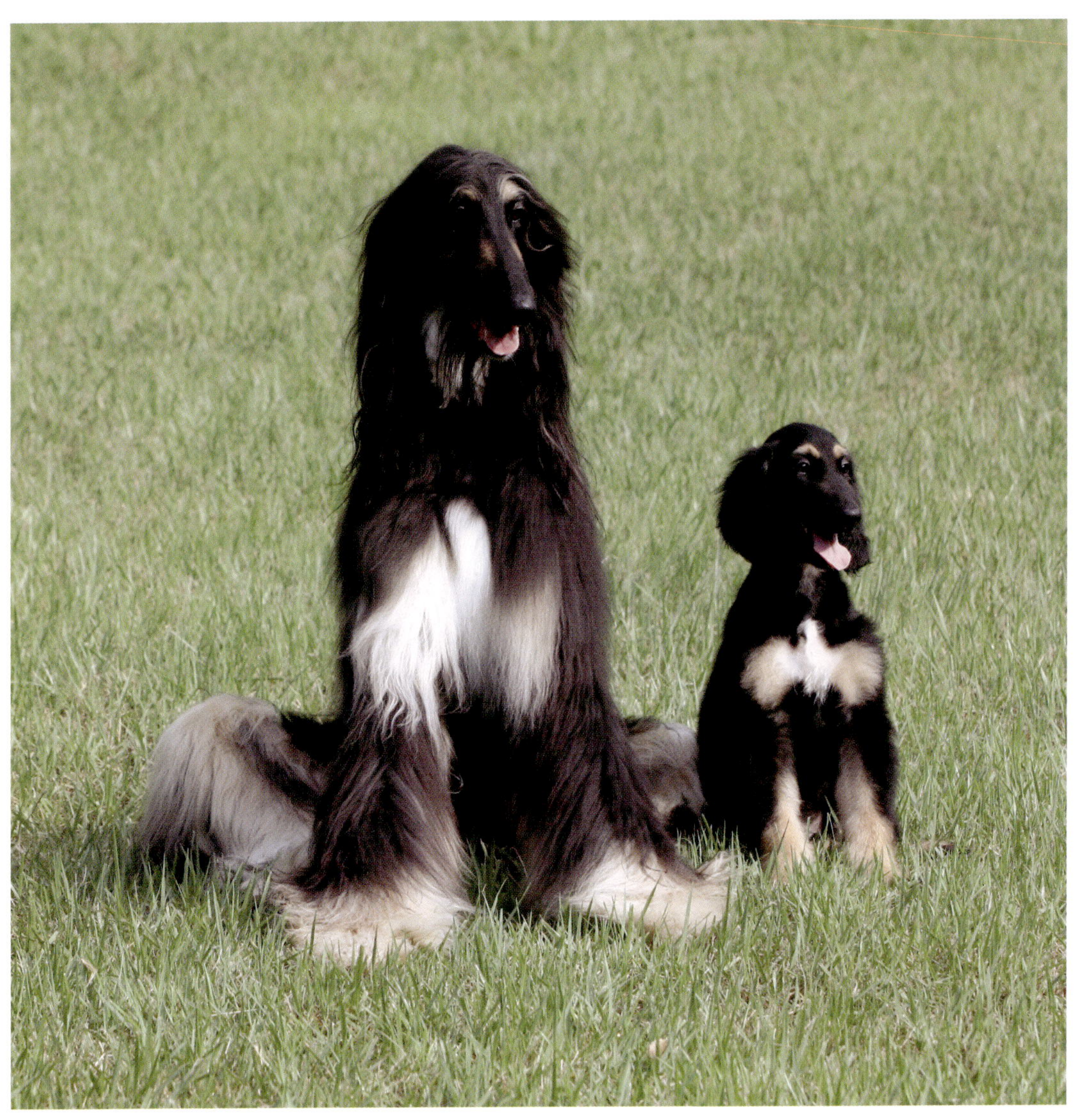

# I AM I BECAUSE MY LITTLE DOG KNOWS ME.

- GERTRUDE STEIN -

I wonder why dogs in the street recognize one another and never mistake me for one of them. Sometimes it makes me sad for I would like to be greeted for what I am, in my friendly-doggy-waggy-arf!-arf! complement. Dogs are often evoked as one of the beings most faithful to man. Is it only the dog that can be faithful to us, absolutely and without judging? Freud says that, unlike humans, dogs know the difference between a friend and enemy, and they don't switch on us.

* * *

Levinas used to recount that in the camp where he was prisoner, in the living form of Bobby, a dog, some humanity was bestowed on him. Marching back to the camp at night after having endured the plight of slave labor, Levinas was suddenly greeted. Bobby came running up to him, recognising him, while the Nazi guard treated him like an abhorred dog.

* * *

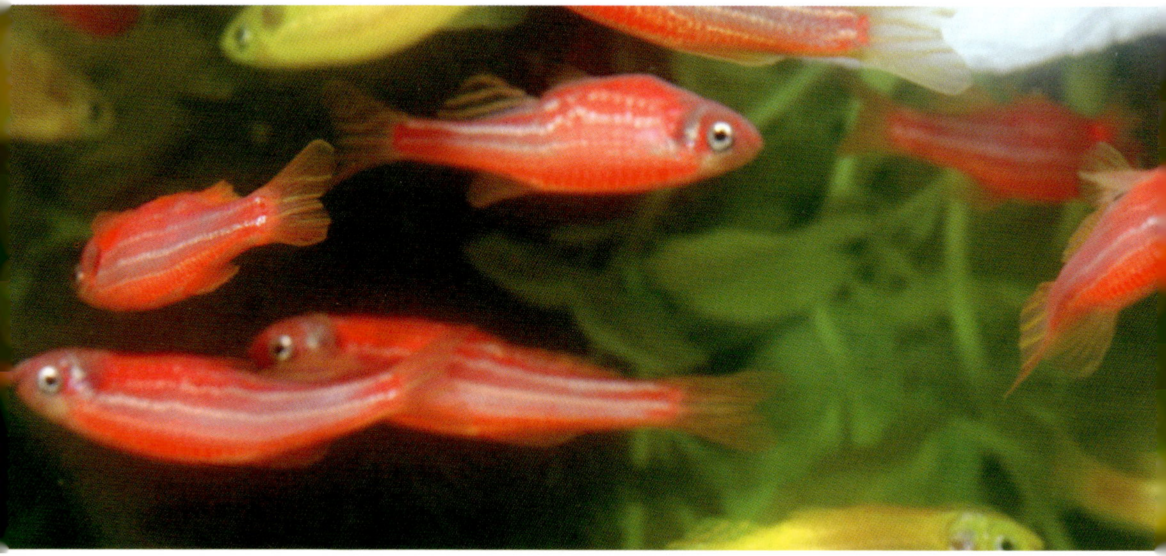

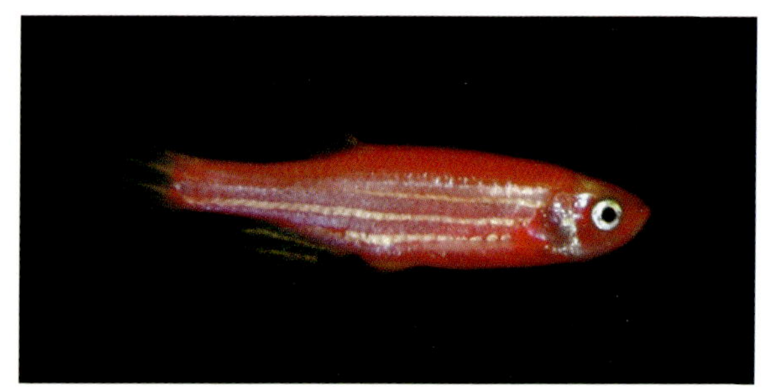

# WE LIVE
# IN OBLIVION OF OUR
# METAMORPHOSES

- PAUL ÊLUARD -

NOTHING IS STABLE.

# IN THE WHOLE UNIVERSE,
# EVERYTHING PASSES;

ALL THE FORMS ARE  MADE ONLY TO COME AND GO.

- OVIDE -

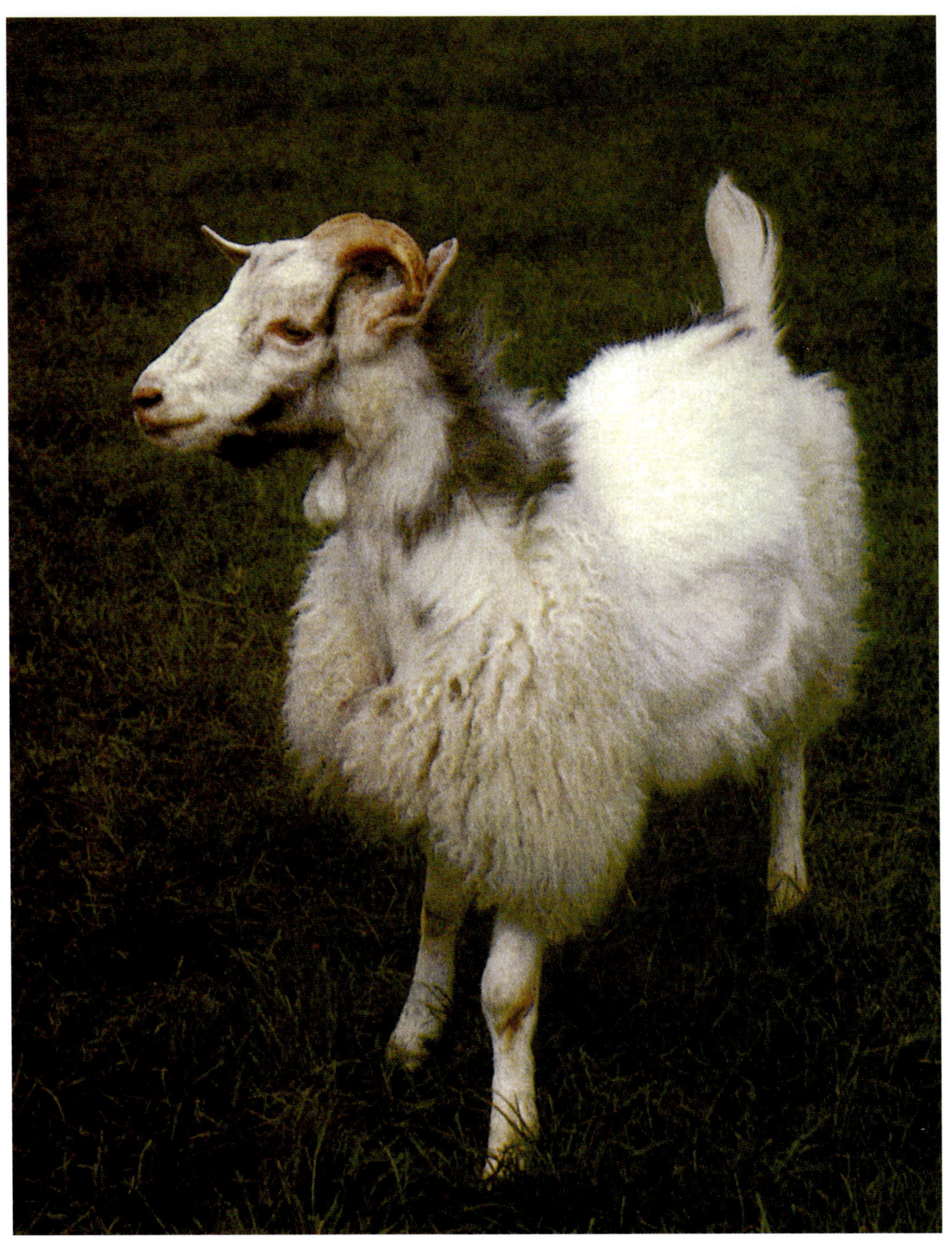

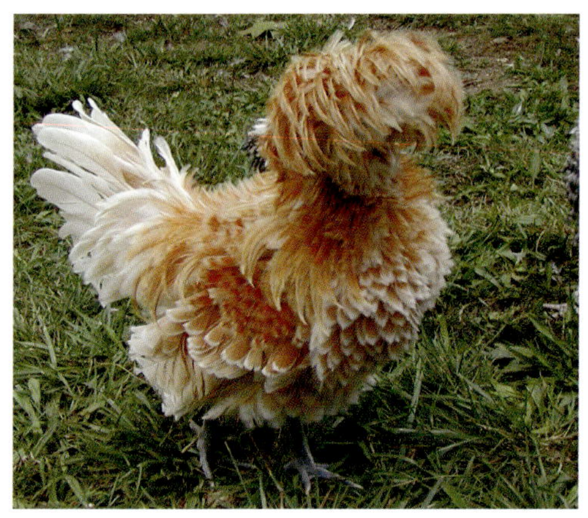

SO THE POET IS TRULY THE FIRE-STEALER.

# HE IS RESPONSIBLE FOR HUMANITY, EVEN FOR THE ANIMALS;

HE WILL HAVE TO INVENT SOMETHING
THAT CAN BE TOUCHED, FELT, HEARD;
IF WHAT HE BRINGS BACK FROM OUT
THERE HAS A FORM, HE GIVES THE FORM;
IF IT IS FORMLESS,
HE GIVES IT NON-FORM.

- RIMBAUD -

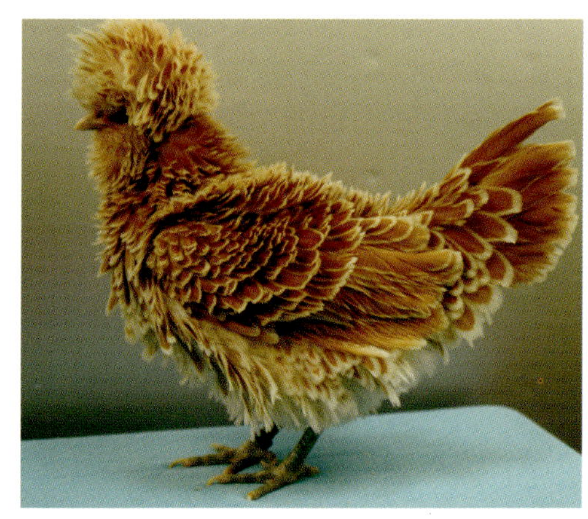

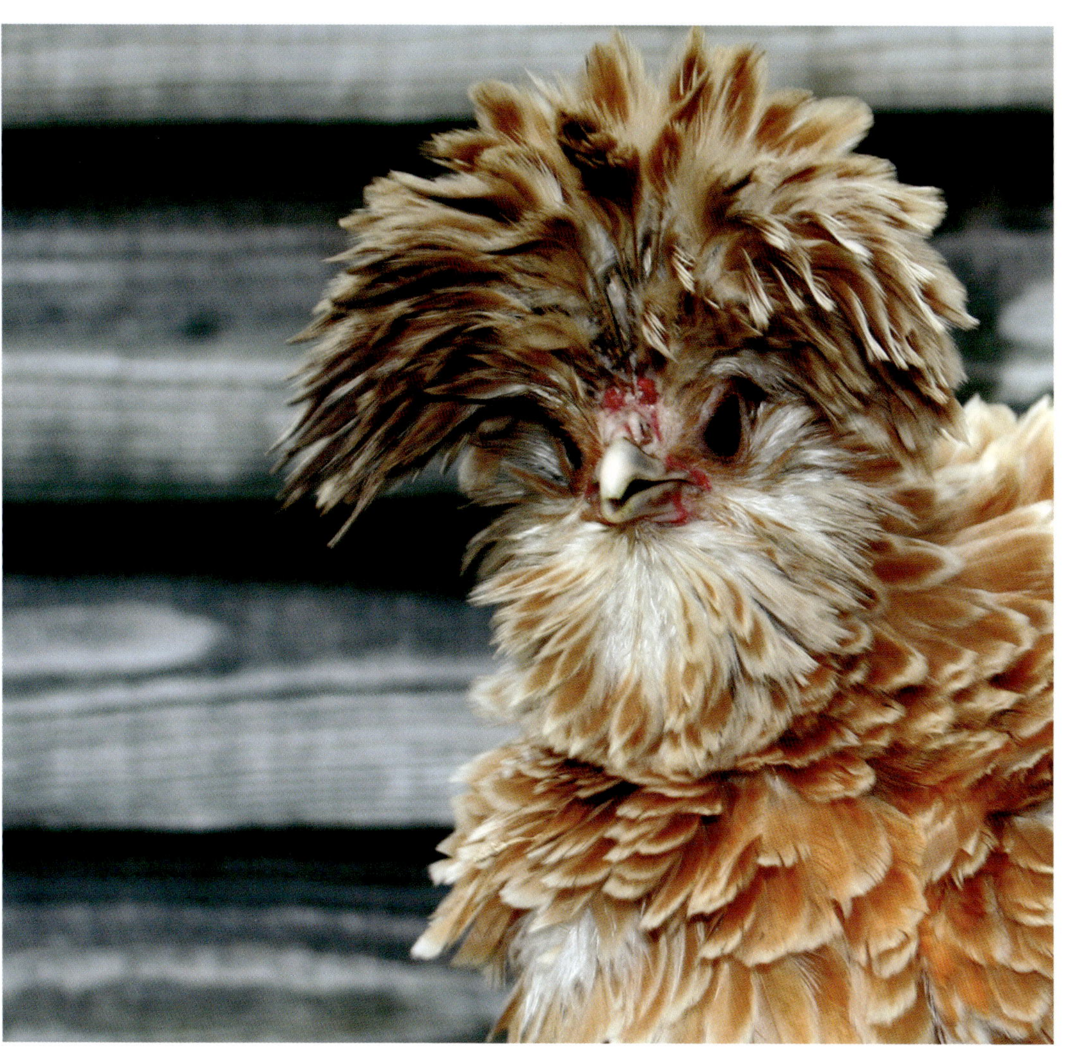

COME BACK TO ME, GONGYLA, HERE TONIGHT,
YOU, MY ROSE, WITH YOUR LYDIAN LYRE.
THERE HOVERS FOREVER AROUND YOU DELIGHT:

# A BEAUTY DESIRED.

- SAPHO -

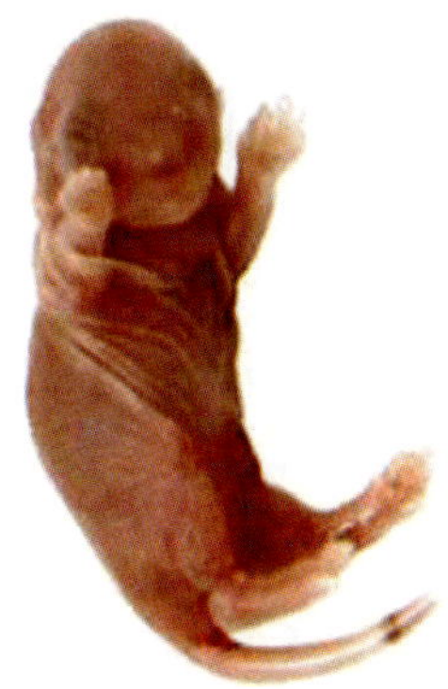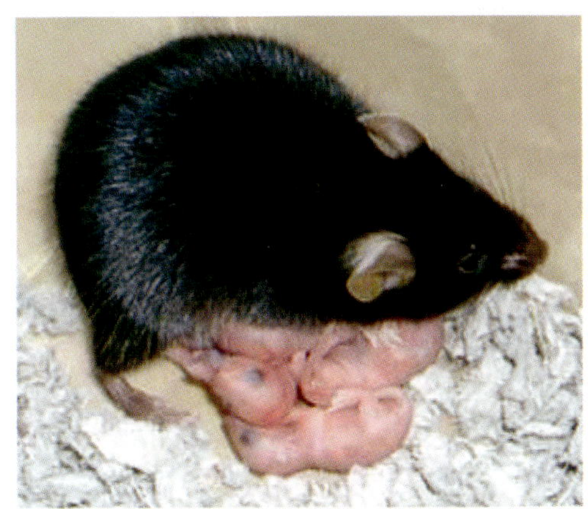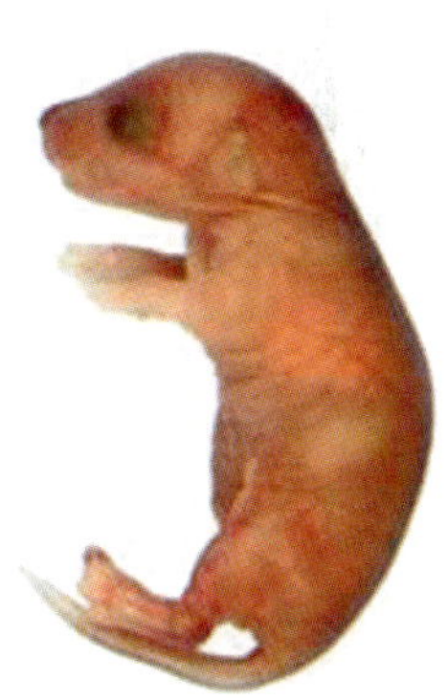

"Do you think that sciences would have taken shape and become great if they had not been preceded by magicians, alchemists, astrologers and witches?", said Nietzsche, describing the exorbitant promises made by these figures out of which the scientist grew. The philosopher used to identify with Eve because she was, according to him, the first scientist who, curious beyond bounds, wanted to go beyond the limits of the human by staying in contact with that snake.

* * *

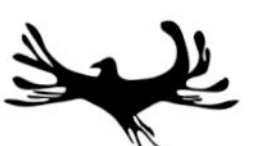

Some will never forgive Valerie Solanas for having shot Andy Warhol. Her target was the male. Why then to have chosen Andy Warhol, who was not entirely the paragon of normed masculinity? He was a hybrid or a techno-zombie who, in a certain way, had already redefined the representations of sexual difference. But she thought that one day one could do without men. They were no longer viable according to the predictions of scientific probity. Technics and science in the service of radical feminism. She placed her bets on hybridization, mutation, and the chromosomal flip out of masculinity.

* * *

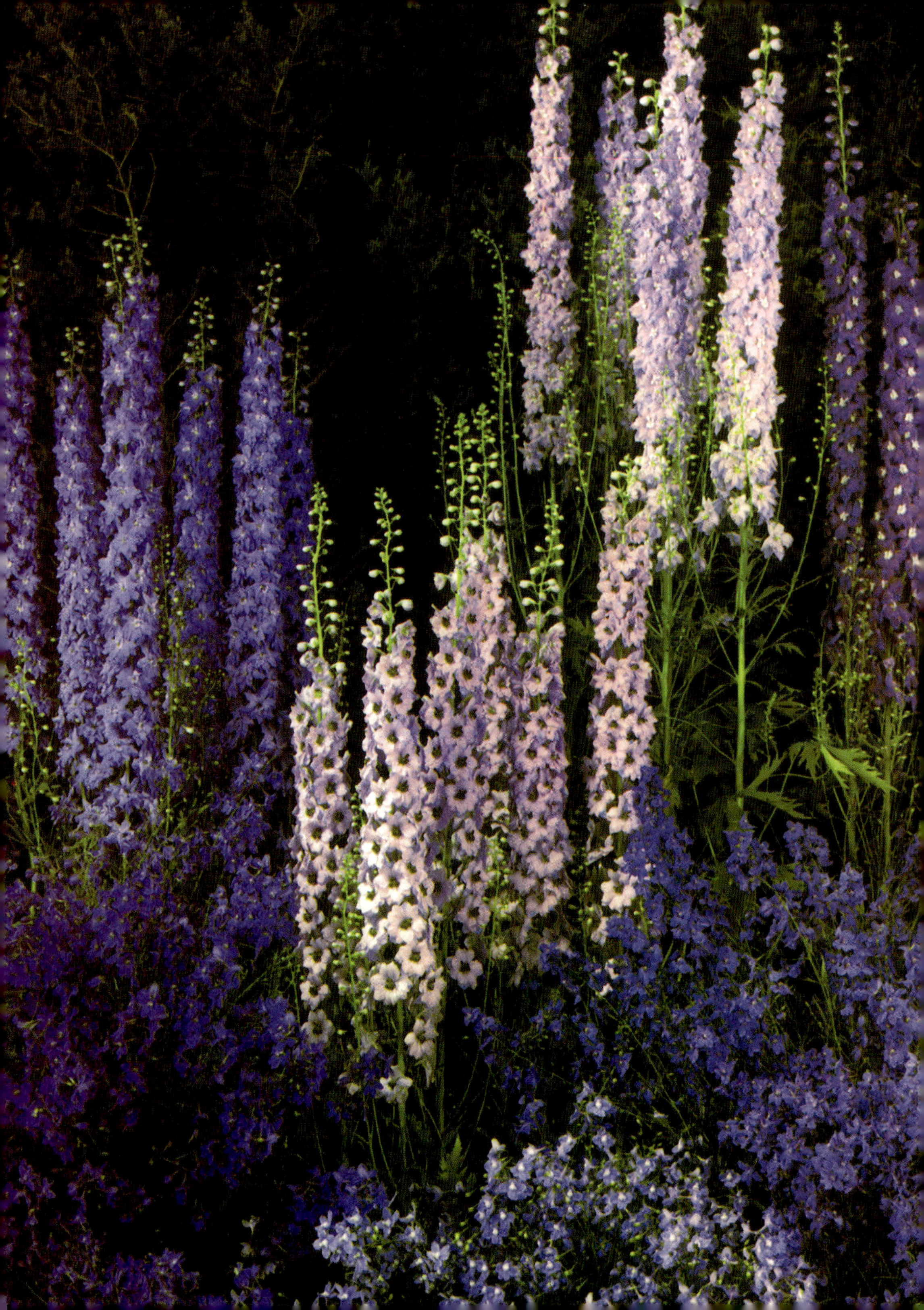

ONE IS A TWO IS A NUMBER.
JOIN THEM AND CROSS THEM
AND SEE THEM BE NUMBERS
BE NUMBERS BEYOND NUMBERS.

TOSS THEM IN WANTON SPIRALS.
WEAVE THEM IN GRAVE COMMUNIONS.
FRAME THEM WITH LIGHTED EYELASHES.
LET THEM HAVE OPENING CLOSING LIPS.
THE WIND IS A WHEN AND A HOW
AND A GIVER OF LAUGHING NUMBERS
AND A THROWER OF CRYING NUMBERS.

ONE DELPHINIUM BY ITSELF
IS A WHO AND A WHO.
A STALK OF BLUE FROM A WEAVING EARTH
A SHEAF OF SKYBLUE FROM A WALTZING SUN
AND ONE IS A TWO IS A NUMBER
AND A SPOKE OF LIGHT IS A WHY
AND ONE YES ONE IS A WHO AND A WHO.

- CARL SANDBURG -

There are poetic influences in science, especially in physics, which looks for extraordinary, original and incomprehensible figurations. And, as well, poetry draws from science.
For Kant, Nature was entirely capable of producing and generating its own "monstrosity" and it is in the excess, notably, that the figure of the poet emerges. He maintains a relation both to the technicity of language and to scenes of Nature, held in the clasp of a "biopoetry". The poet is at the same time that little monster cut from Nature and its aberrant production.

* * *

It is at the time when electricity is discovered that Mary Shelley begins to think about her monster. She is somewhat infatuated with Mr Cross, the great technician of the time, who is interested in electricity, at the same time production of nature and of technology. In her text *Frankenstein: The Modern Prometheus*, she tries to describe the painful story of Victor Frankenstein, the inventor, who stands for a scientist obsessed with the possibility of re-animating his mother's corpse. That laboratory work takes more than two years; a monster comes out of it, that he presents as his thesis for his doctorate at the university. At the beginning of the story, he is seen with his "more than sister" Elizabeth, who incarnates Poetry, and they are inseparable. Then he becomes the scientifically driven Doctor Frankenstein. The monster—the completed thesis—with which his name is evermore associated, is the unnameable. Everything about him is improper. That is where Mary Shelley situates the catastrophe one is talking about, a science that would have radically forgotten poetry: science will have obliterated its origins and complicities in poetic reflection. And that is what Mary Shelley cried over, the rupture between science—which has become intricated in technology—and art that she builds on sexual difference. At the end of the story the monster commits suicide. It has learned what life is and what finitude must be by reading *The Sorrows of Young Werther*. As with Goethe, too violent a separation between art and science culminates in a call for death, a death-work.

* * *

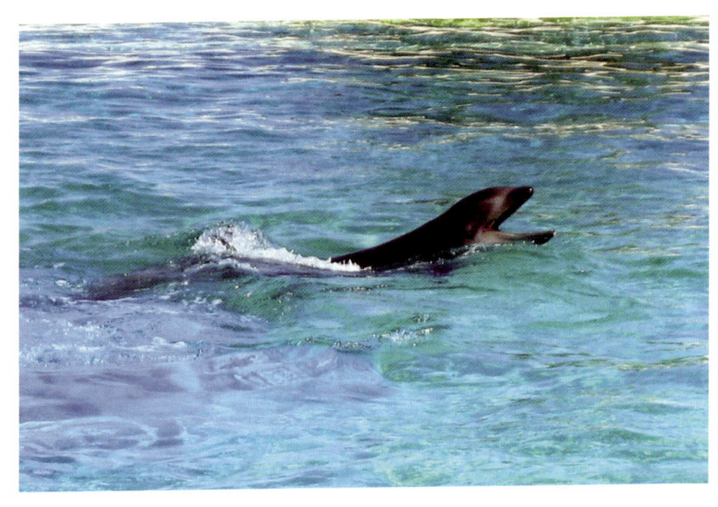

# AN ANIMAL'S EYES HAVE THE POWER TO SPEAK A GREAT LANGUAGE.

- MARTIN BUBER -

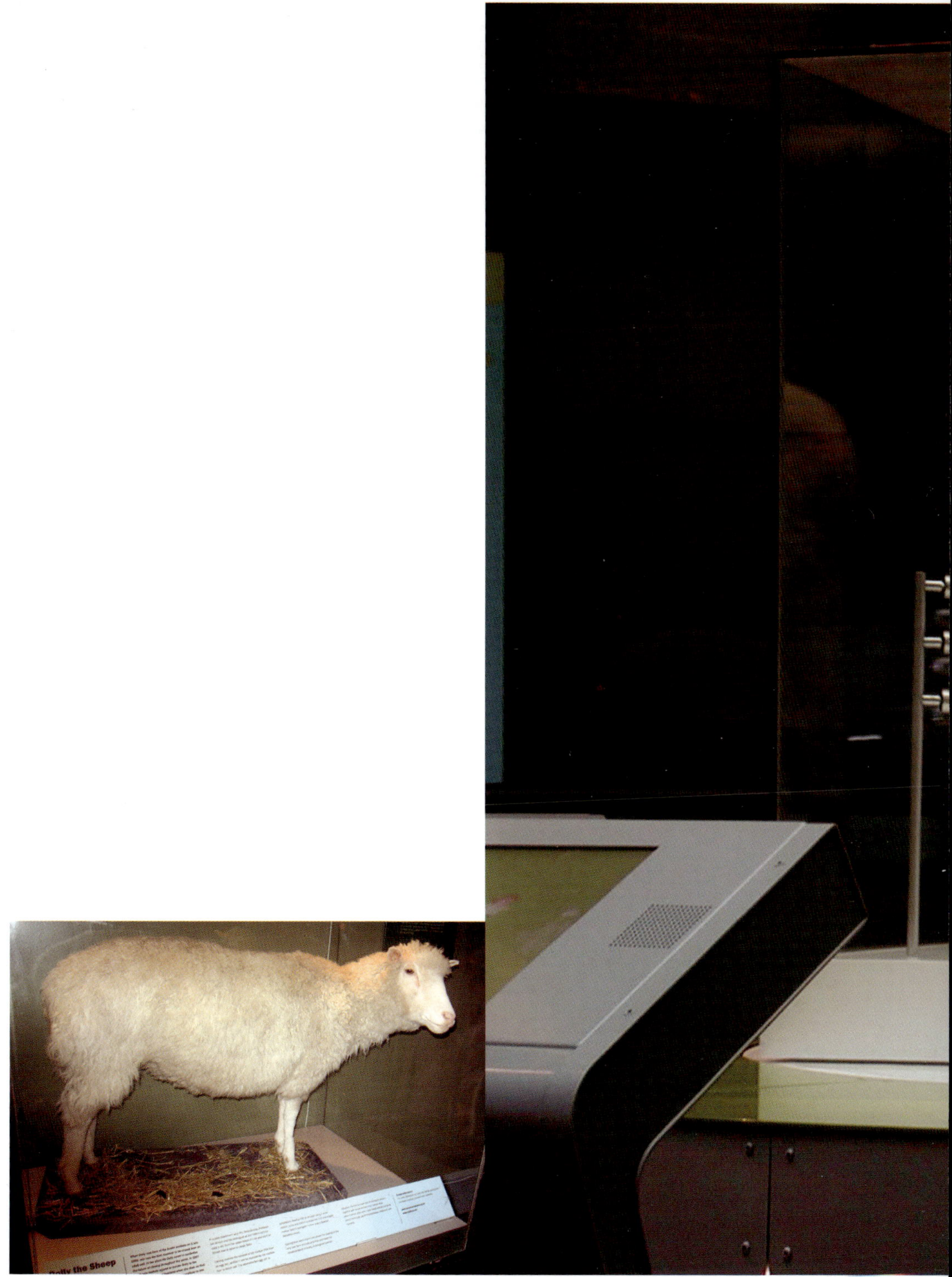

Dolly the Sheep

In *Blade Runner*, the question is to know what it is to be a human being. The "replicant", an artificial being created by the Tyrell Corporation, tries to pass for a human being. But it fails: the test to which it submits itself doesn't acknowledge the mother. To be a human is to have posited an origin, a memory, and to have been born by a maternal incubator of sorts. Properties that the replicant-cyborg, the extended "robot," can't pretend to have. It has been manufactured solely to obey and satisfy human needs. By inviting animals to "speak", Jacques Derrida tried, perhaps, to tackle that so-called supremacy.

* * *

What is Nature? What are the aberrations, the oversights, the fissuring that constitute the presumed natural setting? Even in a laboratory, nothing is natural according to contemporary German scientist H-J Rheinberger, who tries to establish a grammatology of that place where everything is framed, where everything is already artificial, technical. "Nature" is no longer even when one thinks one can look into so-called natural objects. One could also ponder on that need to test everything. It is important to bear in mind that the bureaucracy of humiliation was based on the ideology of the scientific test—everything and everyone was submitted to testing, one could neither be admitted to or exit institutions without passing a test.

* * *

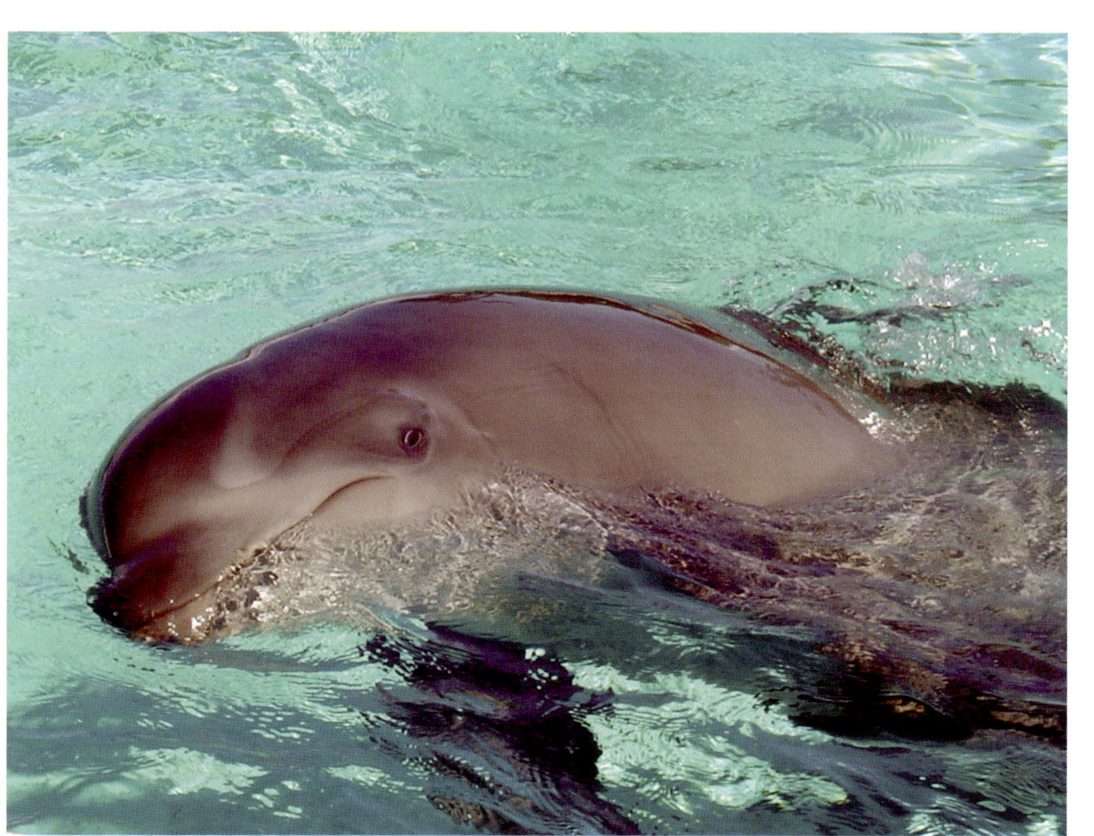

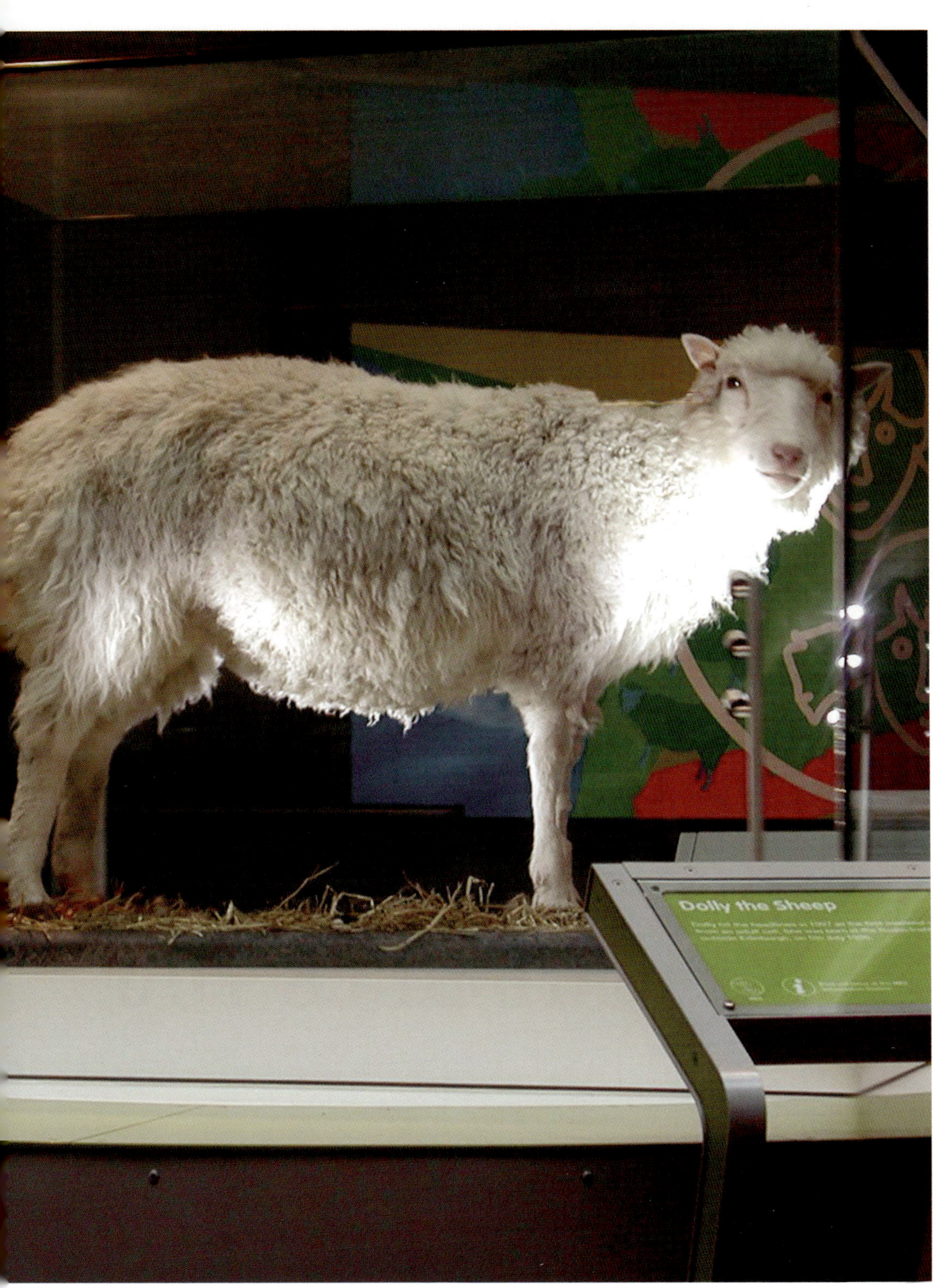
Dolly the Sheep

# FOR OVER THIRTY YEARS NOW I HAVE BEEN INDULGING IN THE OCCUPATION OF RAISING GOOSEFLESH DOMESTICALLY AND IN ENGLAND.

(...) NATURALLY, MOST OF THE ACTORS IN *THE BIRDS* ARE. IN FACT I HAVE EMPLOYED MORE FEATHERED PERFORMERS THAN HAVE BEEN SEEN SINCE PAN DANCING WENT OUT OF STYLE.
THE CAST ALSO INCLUDES A FEW MEN AND WOMEN.

# AFTER ALL, THE PICTURE ISN'T JUST ABOUT THE BIRDS, IT'S ABOUT THE BEES TOO.

- HITCHCOCK -

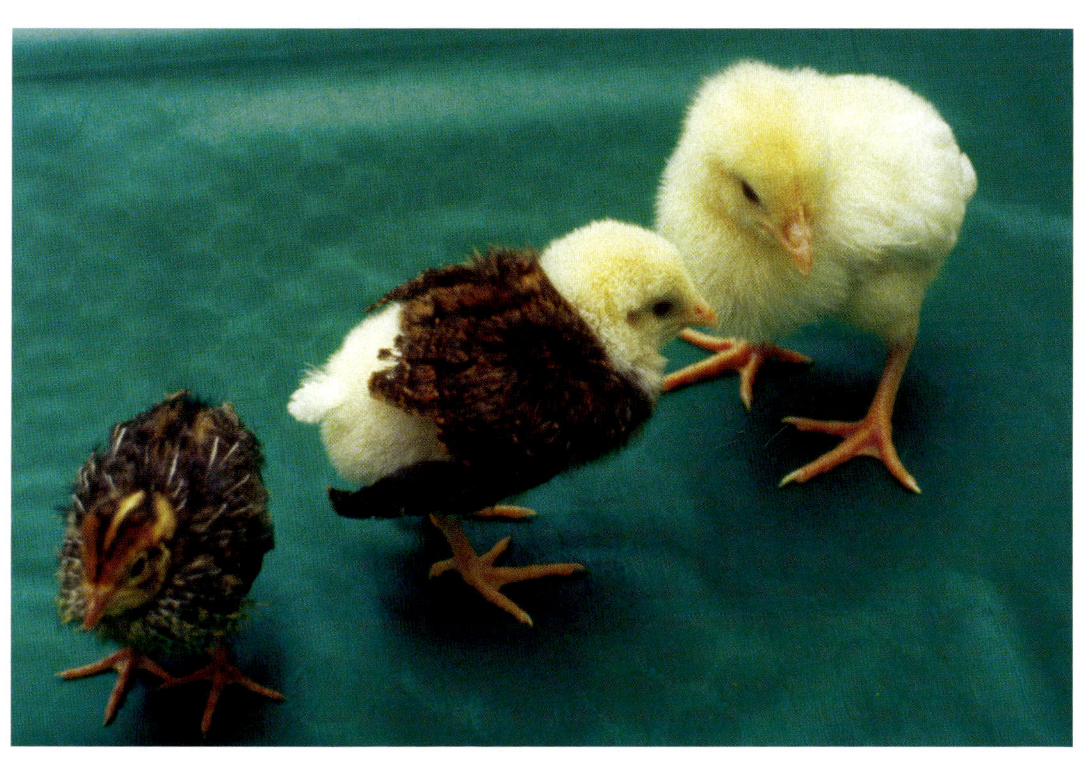

For quite a long while, like Hitchcock, I've considered birds as a kind of meta-psychological feat. The bird is never what it is empirically, it surpasses itself, it is already spiritualized, inscribed as something which charms and threatens, which can kill or save. Eckermann, who was Goethe's secretary, helped him to do his research on the mutations of colours. He was obsessed with birds. He lived in a little house behind Goethe's, with forty specimens, some of which were predators. Try to imagine forty birds, it's a lot in a small space. After Goethe's death, he finished his work during the night, under dictation. Alone, he carried on listening to him, taking notes to finish that book they have started together. The birds even shat on the papers. It will become "Conversations with Goethe" that Nietzsche and others considered the most beautiful book in German letters. Some even pretend that Goethe in person signed that posthumous work and that Eckermann was nothing but a dictaphone.

* * *

Passing by Goethe, Hitchcock and others, I'm trying to say, quickly, that those beings have never been able to simply remain what they are. They are like synecdoches, less than themselves and more than they appear to be.
Birds fly in very phantasmatic spheres, and are not only "edible". What does one incorporate when one eats the other, when one downs a bird? This was the question posed by Hitchcock in the detachable introduction to the film, when, speaking of the meaning of the work, he shoves away his plate of chicken.

* * *

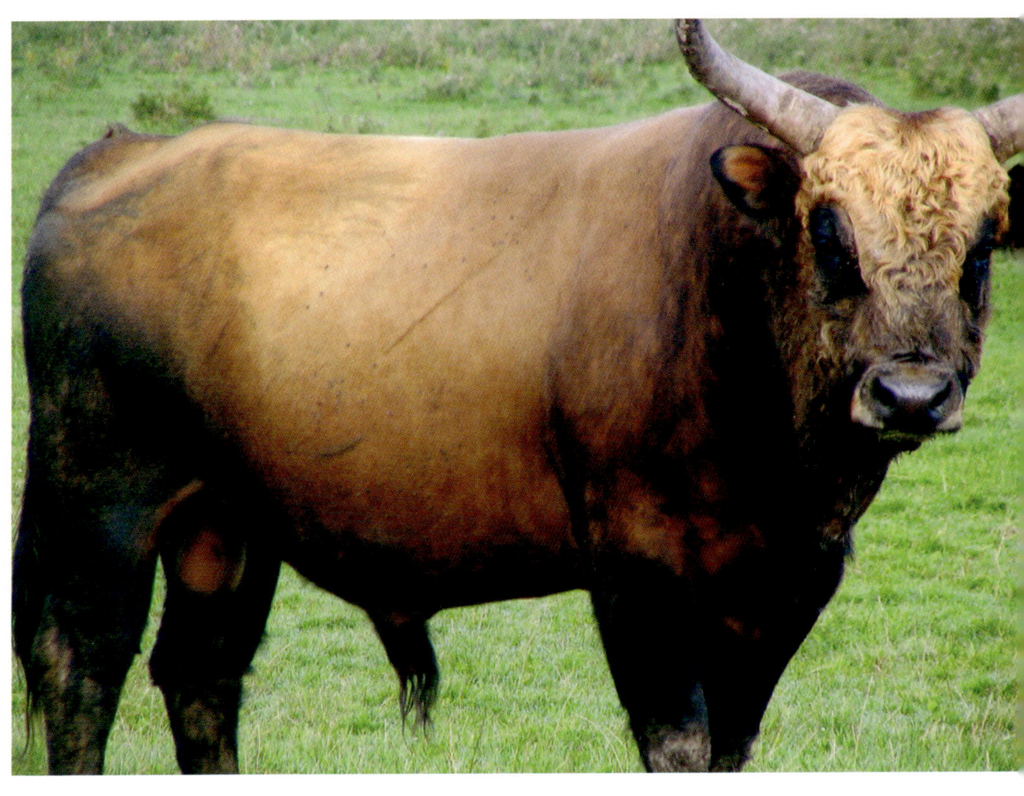

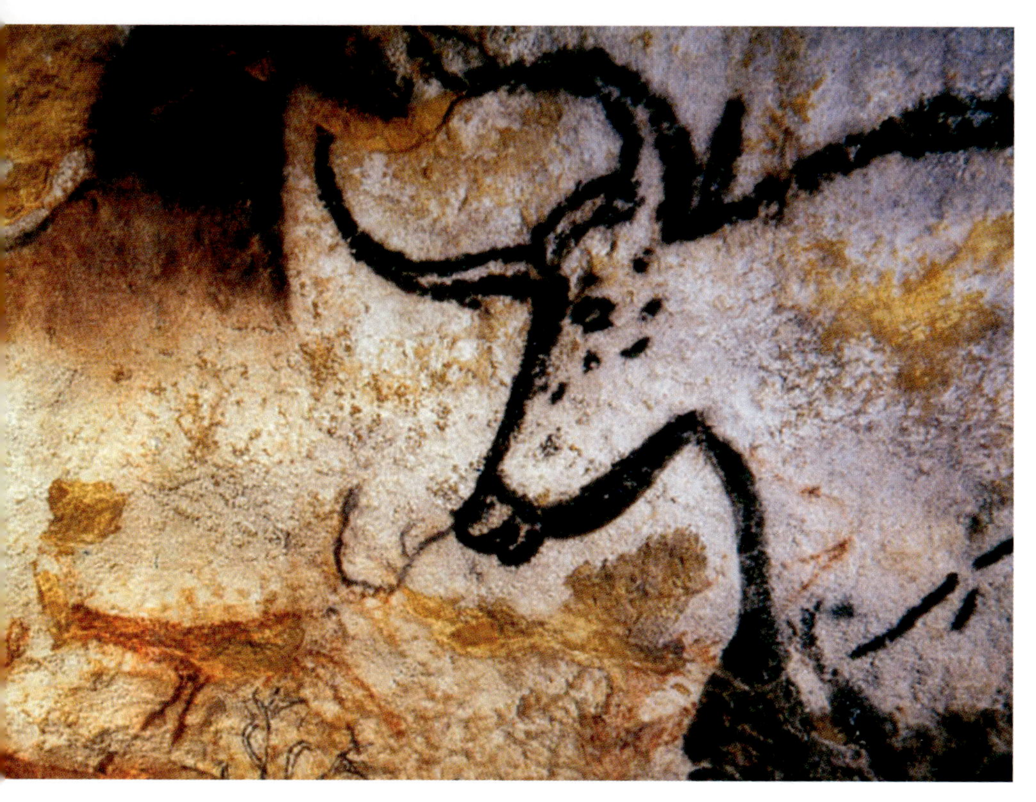

# IT'S EVIDENT

THE ART OF LOSING'S NOT TOO HARD TO MASTER

## THOUGH IT MAY LOOK LIKE

# (WRITE IT!) LIKE DISASTER.

- ELIZABETH BISHOP -

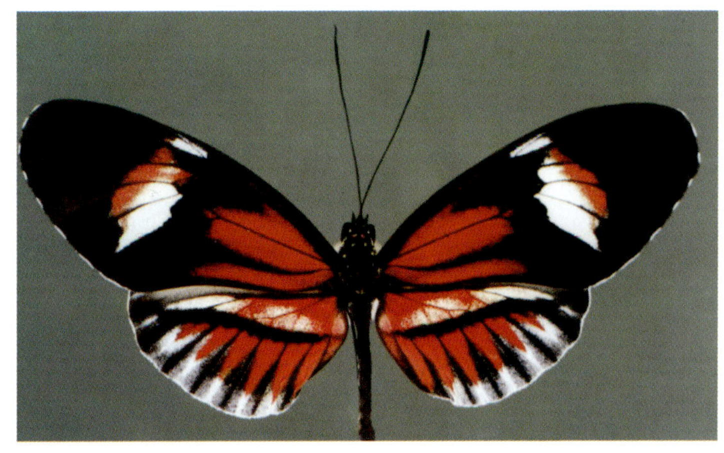

**ROUTINE IS AN ILLUSION
AND CAN BE REPLACED BY
THE DISCOVERY OF NEW PHENOMENA.**

- FLÁVIO DE CARVALHO -

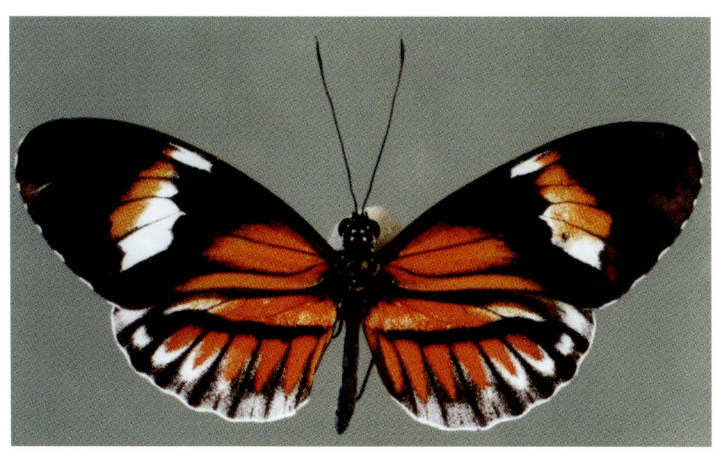

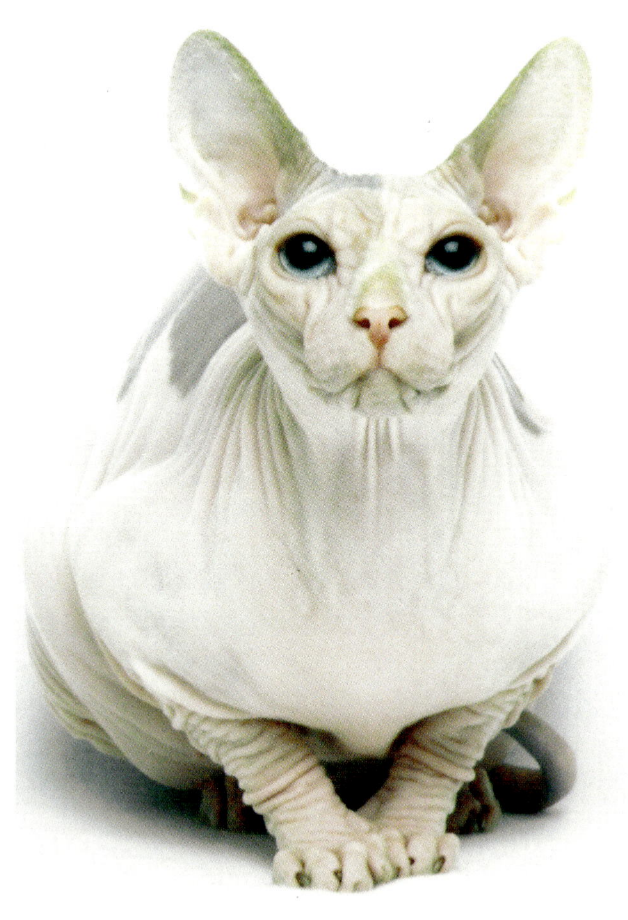

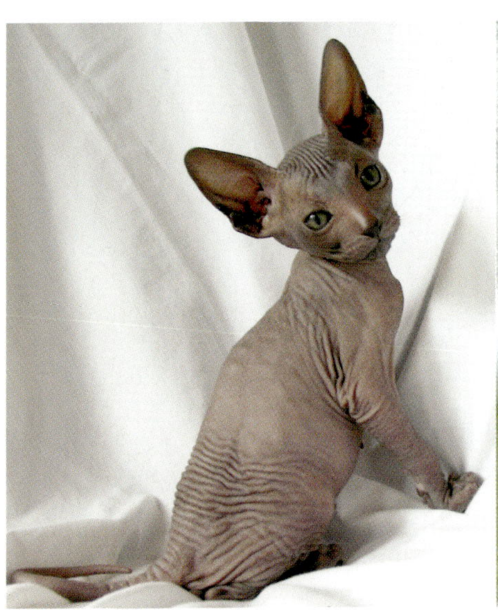

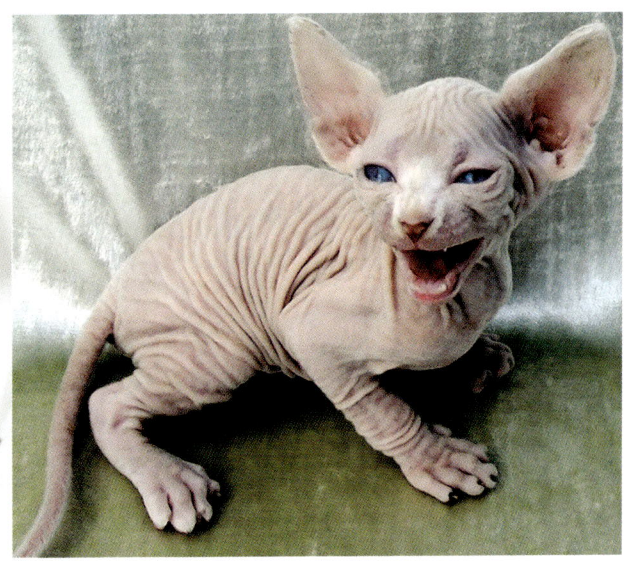

WHICH CREATURE
IN THE MORNING GOES ON
FOUR FEET, AT NOON ON TWO,
AND IN THE EVENING
UPON THREE?
SPHINX

MAN—HE CRAWLS
ON ALL FOURS AS A BABY,
THEN WALKS ON TWO FEET
AS AN ADULT, AND WALKS
WITH A CANE IN OLD AGE.
OEDIPUS

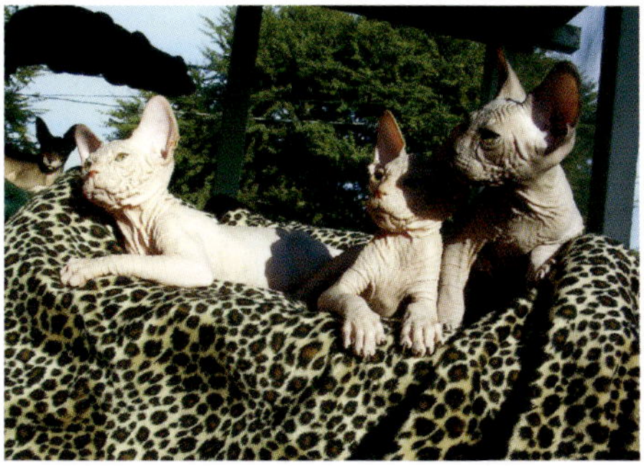

Why does the question of domestication become disturbing now, or even necessary when one is in a world of the technological grid, which, in a way, has uprooted the house and undermined the metaphysics of any household? What is a house today if not a place stuffed with wires where I talk to invisible people, to absent beings. A house can be understood as a phone number, an address, a destination. Often enough, it only becomes a house because an animal lives in it. Is the house today only an opening or access code, traversed by networks, or is it still a place that shelters a tribe or a family?

* * *

What is a family? What is a couple? What is domestication? What is it to live with another person, to domesticate another person, to have another person living in one's house or to create a home with another person? It has to do with that violence of adaptation, of "training" which is not limited to animals… I do to the other what one does in the cattery with the sphinx. I try to create people who can live with me.

* * *

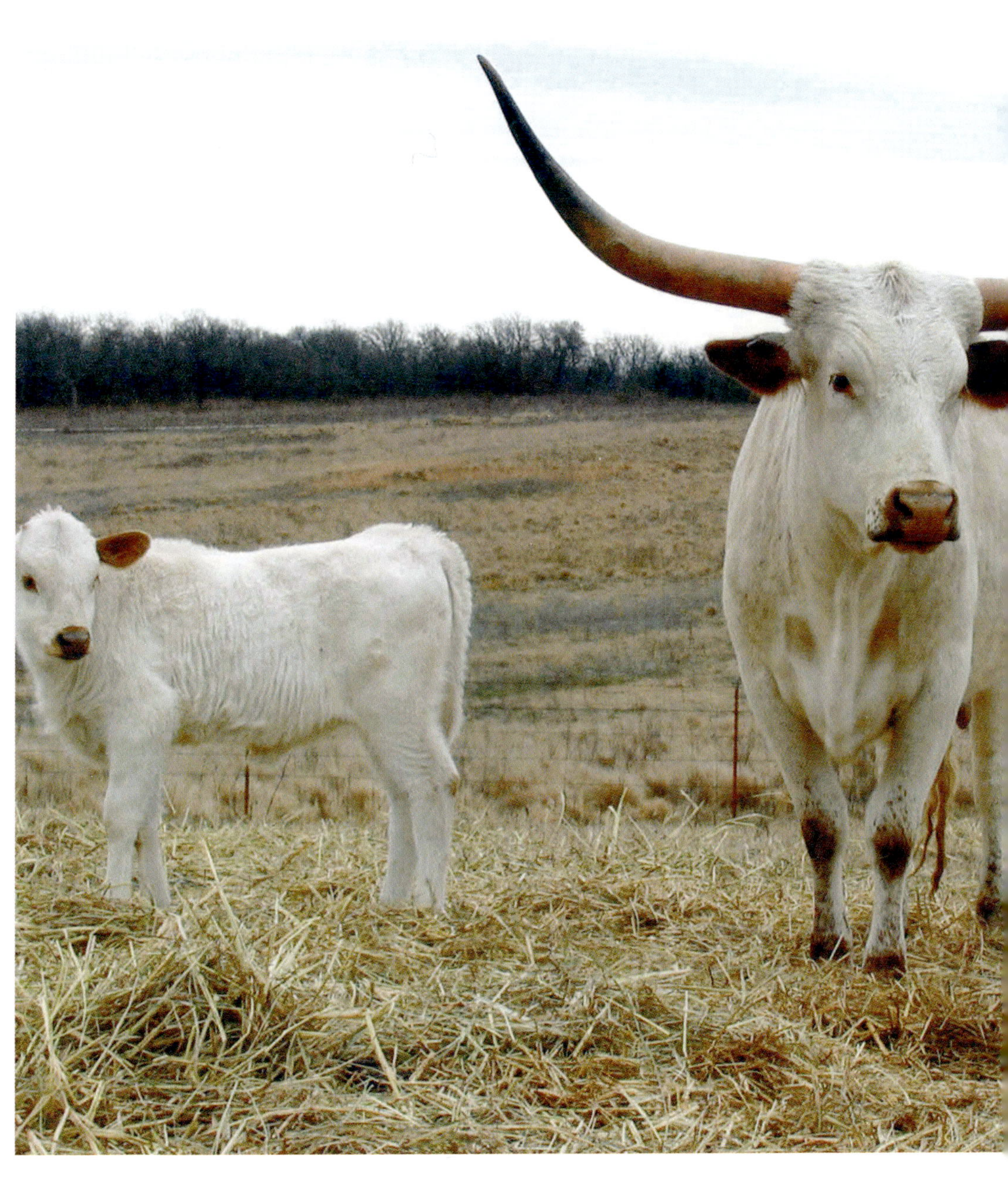

# PERHAPPINESS

- PAULO LEMINSKI -

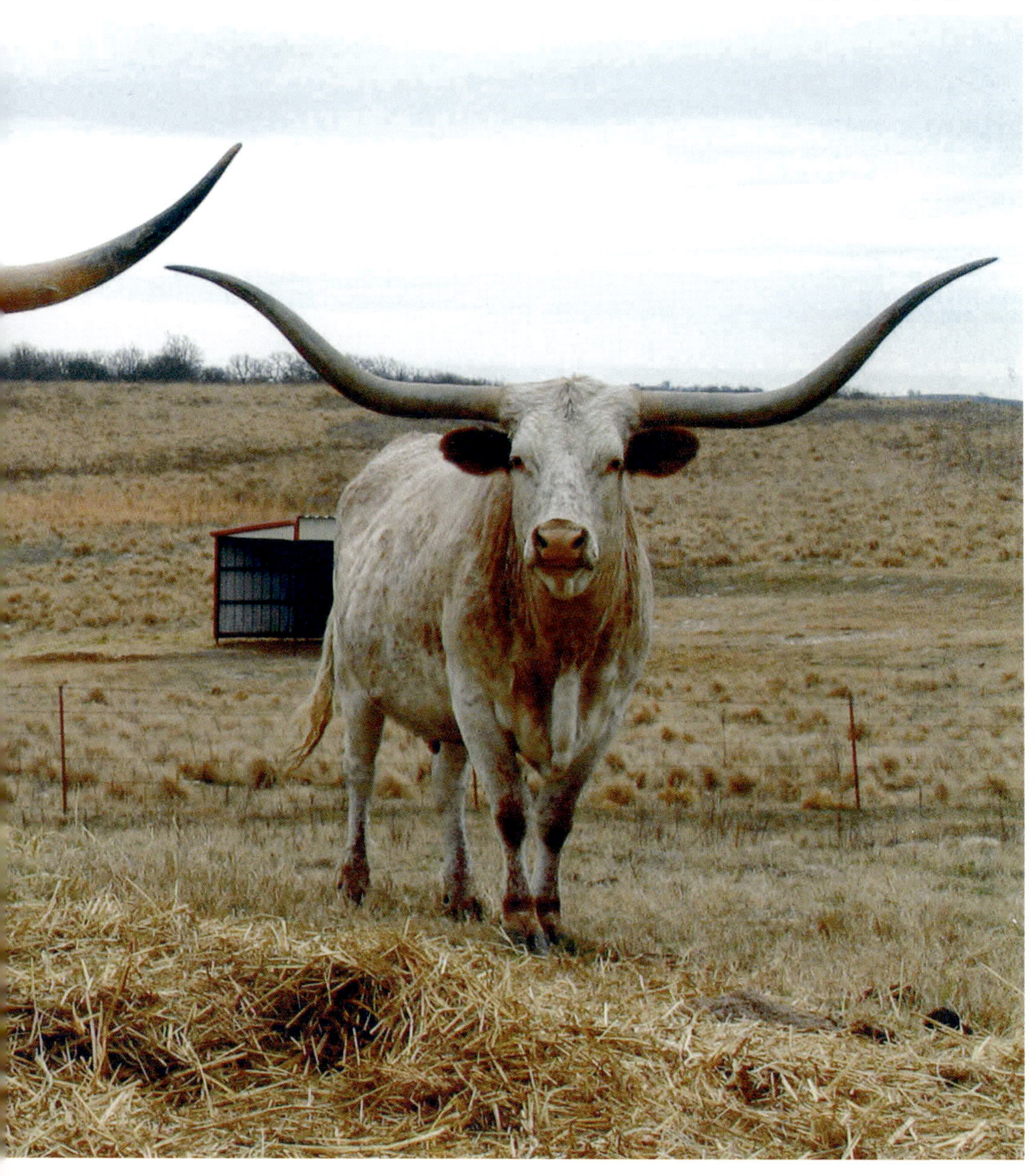

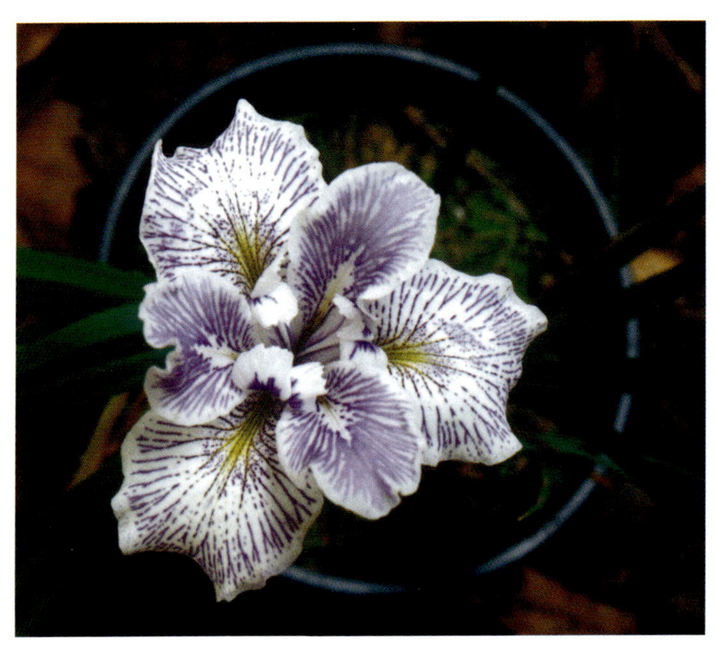

# FOR THOSE [IRISES] THAT ENCHANT ME I WILL BE A PROTECTOR AND A BUMBLE BEE.

- GEORGE GESSERT -

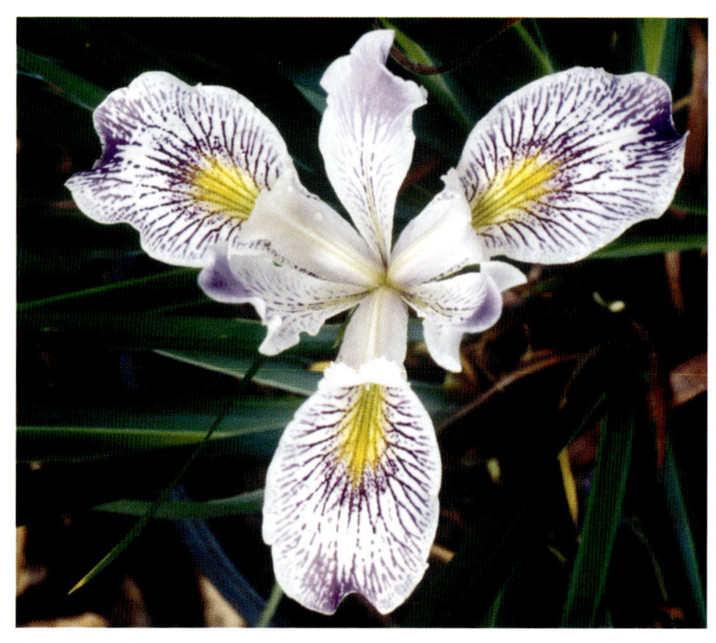

Why should it be unbearable to tolerate a sentence like "we are already transgenetic" or, the opposite, to wonder "in what way does it comfort or reassure us to think that we are already transgenetic"? What is that desire to connect us with our "colleagues," the banana and mustard?

That resistance or impossibility to admit that there are radically heterogeneous parts that cannot be made to fuse or coincide with their concept compels us to rethink History and the concept of the human itself, or to become schizophrenic in order to assimilate it.

* * *

"If we were masters of biology, we would be Gods," said Descartes, who was extremely interested in that idea of living a long life for, in his time, one died young. Perhaps according to future calibrations we all will have died too young, no matter at what age—the age of Descartes and beyond. There are people today who think that we are similar to plants and thus that we should be able to live forever, to hibernate like trees and reblossom seasonally. Empedocles said that man came out of the ground like spinach.

* * *

# MY COMPANION MUST BE OF THE SAME DEFECTS.

# THIS BEING YOU MUST CREATE...

- MARY SHELLEY -

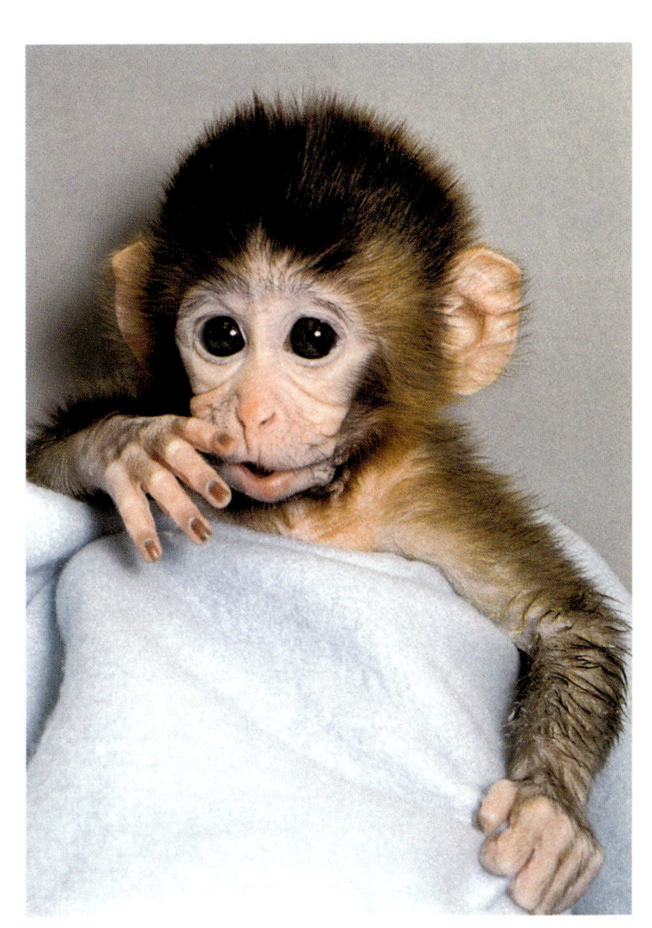

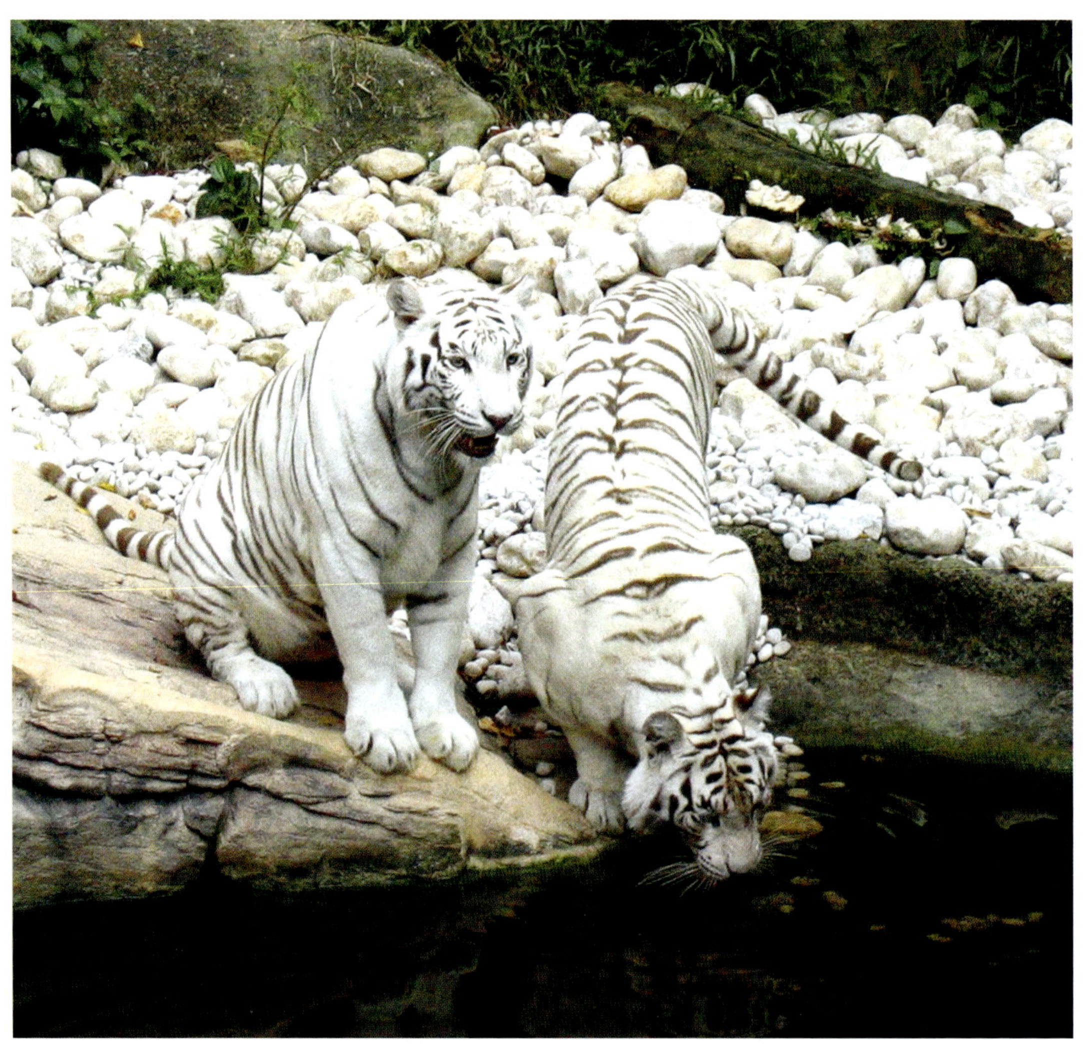

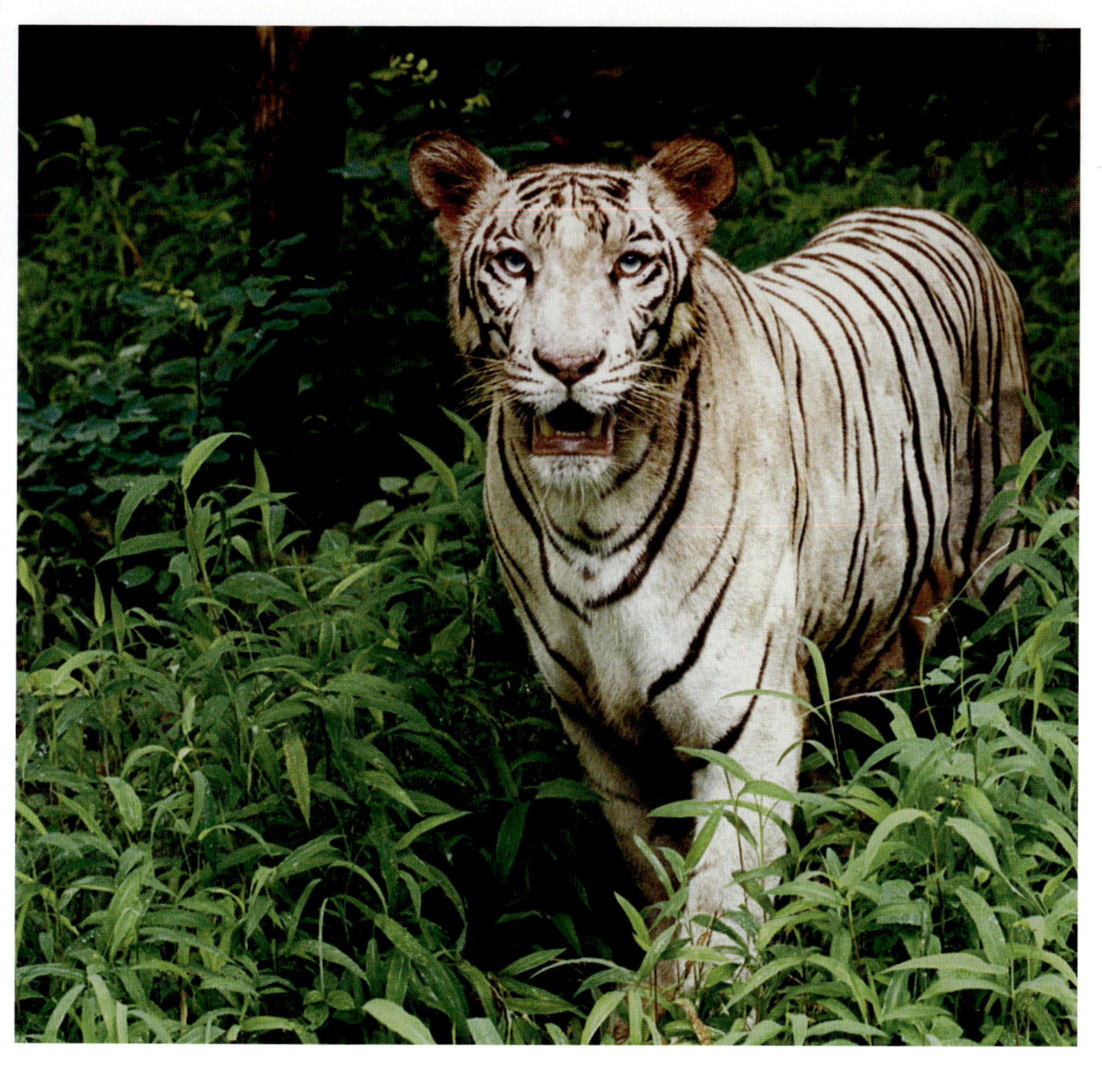

**GENTLY**
    **(VERY WHITENESS:ABSOLUTE PEACE,**
**NEVER IMAGINABLE MYSTERY)**
               **DESCEND**

- CUMMINGS -

# MAKE A POEM
### THE WAY NATURE
# MAKES A TREE

- HUIDOBRO -

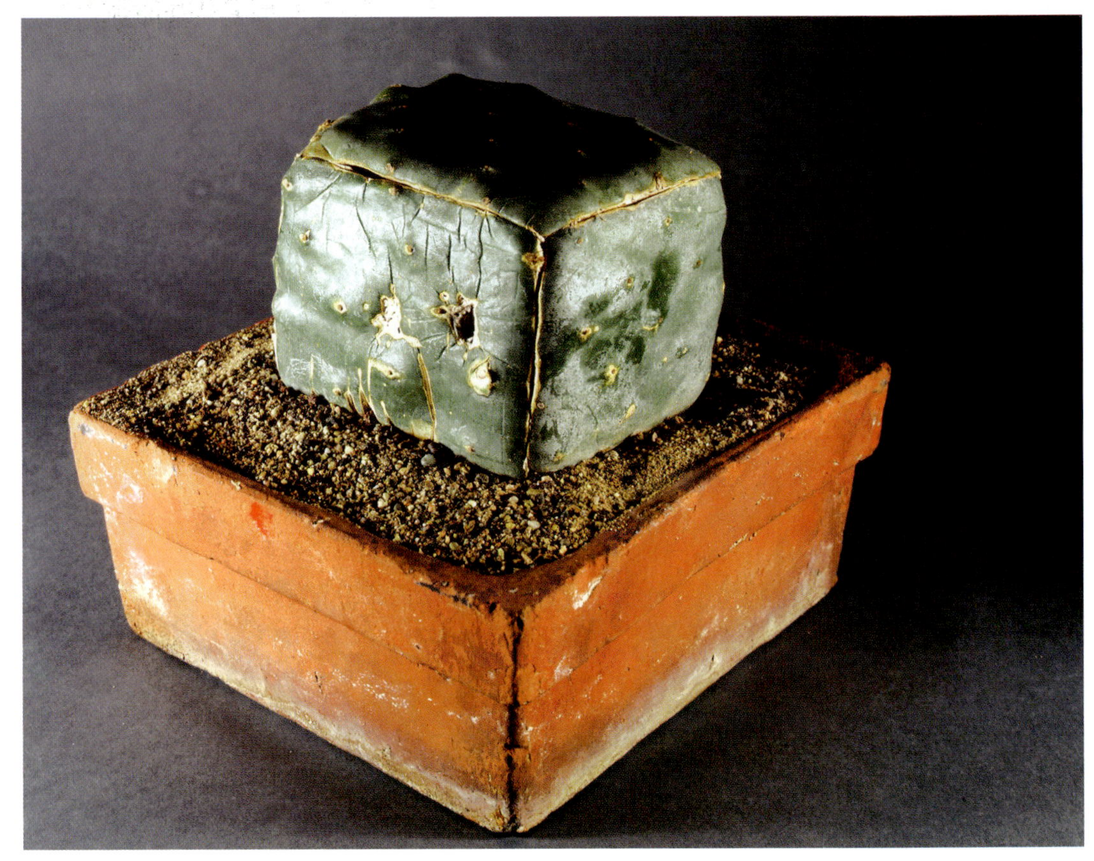

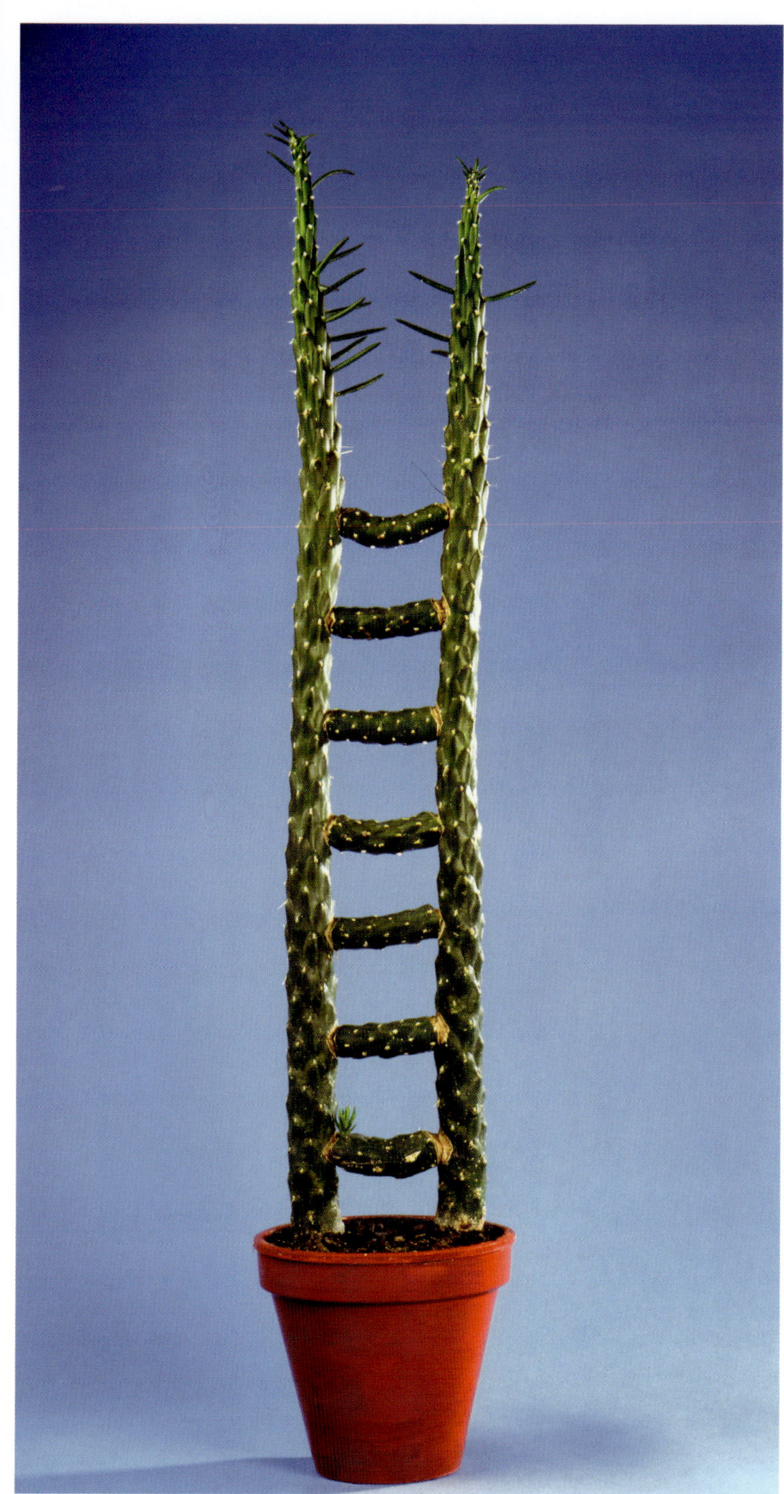

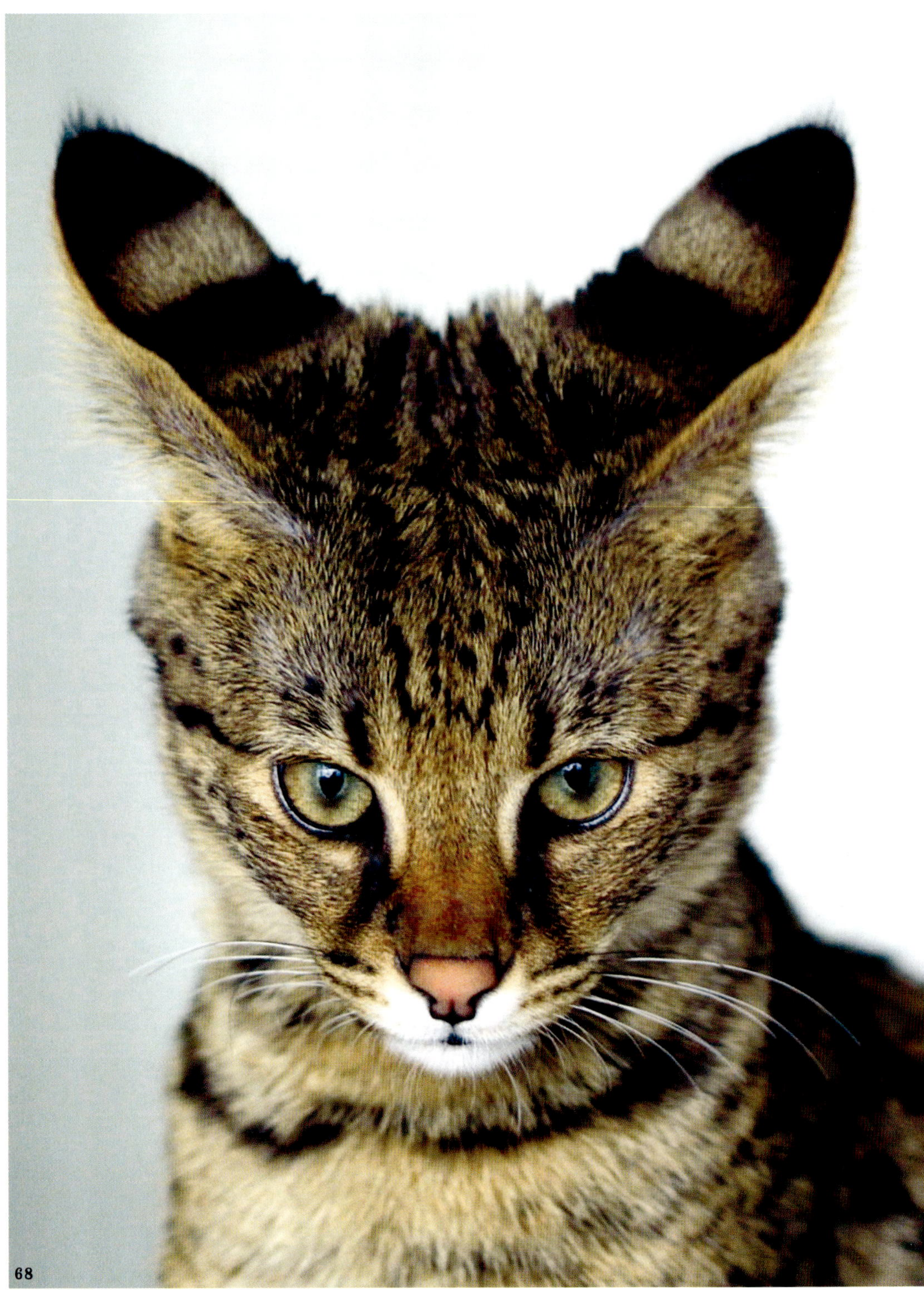

# WHEN I

**PLAY WITH MY CAT,
WHO KNOWS WHETHER
SHE IS NOT AMUSING HERSELF
WITH ME MORE THAN I WITH HER.**

- MONTAIGNE -

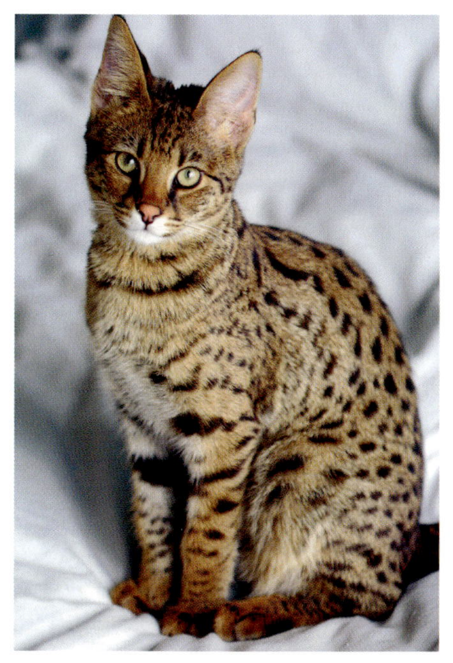

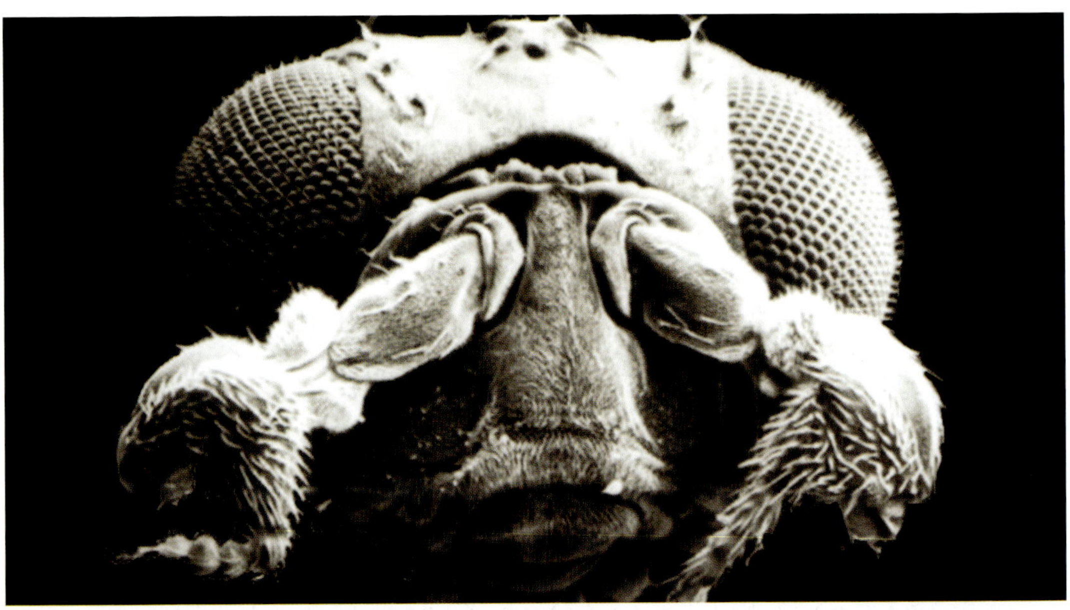

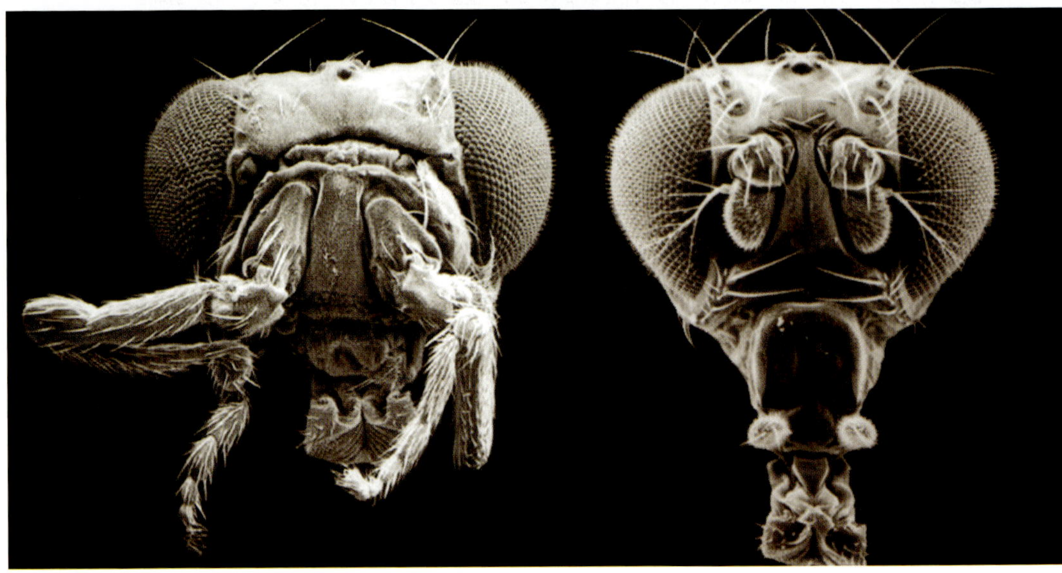

# WE
## SHOULD VENTURE ON THE STUDY OF
## EVERY KIND OF ANIMAL
## WITHOUT DISTASTE;
### FOR EACH AND ALL
### WILL REVEAL TO US SOMETHING NATURAL
## AND SOMETHING BEAUTIFUL.

- ARISTOTLE -

In the past there were gods who roamed our world, who metamorphosed, who raped through deception, and took on dissimulating forms. There were demi-gods, very important for Hölderlin, precisely because they were hybrids who, at the same time, had to pursue an earthly destiny, according to the laws of temporality, and who possessed divine qualities, thus, eternal ones. There are days when something comes close whose features are not readable: as Rilke said, if another being, perhaps an angel, were to manifest itself, I would perish because another "Dasein" appearing, I would run the risk of being annihilated.

* * *

In an old episode of *Star Trek*, beings coming from the in-between space,
themselves between-species, go back in time to decipher the signals they have
received from Earth. They suppose they are human signals. But, like us, they are on
a false track throughout the whole film. In the end, we discover that they are
distress signals from a whale of an extinct species. It is a kind of poem on the
hubris, on the arrogance of human beings—who think they are the only recipients
but also the origin of all readable and intelligible signals, of the whole store of
meaning's excess—and that I saw as post-human.

* * *

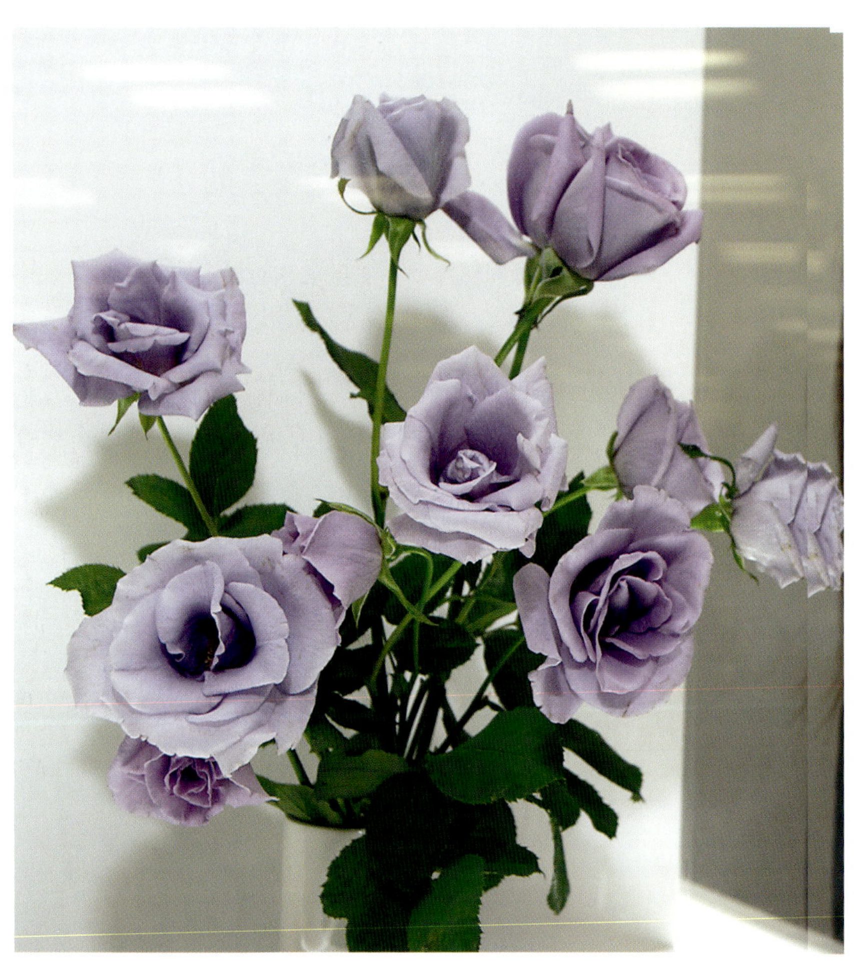

IT'S NOT THE TREASURES I CARE
ABOUT, HE SAID TO HIMSELF,

SUCH COVETING IS
MILES FROM MY MIND,
BUT I LONG TO SEE THE
BLUE FLOWER.
I CAN'T GET RID OF THE
IDEA, IT HAUNTS ME.

- NOVALIS -

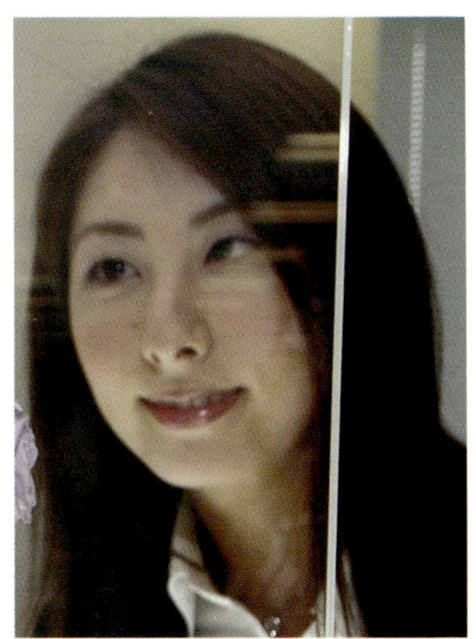

The blue flower didn't exist in the referential world, only poetry could imagine it.
There have been protagonist figures searching for those non-objects, those imaginary
things, as in the novels of Novalis. Today everything can pass into existence,
everything that was once limited or everything that lived in the imaginary can now
have an empirical or at least a closer to real life..

* * *

There is an ethics of limits, a thinking of the limit, which does not coincide with
closure or ending. Be that as it may, one always assumed that one hadn't yet gone
too far, that we are still well within the allotted limit of doing and saying and being.
I, on the other hand, wonder if one has gone too far in our time. What is "far" or "too
far" or, for that matter, "our time"? One should try to pin-point what the excess of
technological accomplishment implies. Despite a kind of jubilation of the possible
that often surrounds the technological element in art or warfare, it is feasible that one
has transgressed the limits of human destructiveness. These are questions that recur,
for example, concerning the relationship between the limit and the end. Melanie
Klein says that there are often fake repairs or manic reparations, which hide the fact
that not everything can be repaired. There are things that are definitely erased,
irreparable. Have we gone into the zones of the irreparability of our existence and our
history? It may be that we have gone too far.

* * *

# THERE IS NO REWIND BUTTON ON THE BETAMAX OF LIFE.

# AN IMPORTANT EVENT TAKES PLACE ONLY ONCE.

- NAM JUNE PAIK -

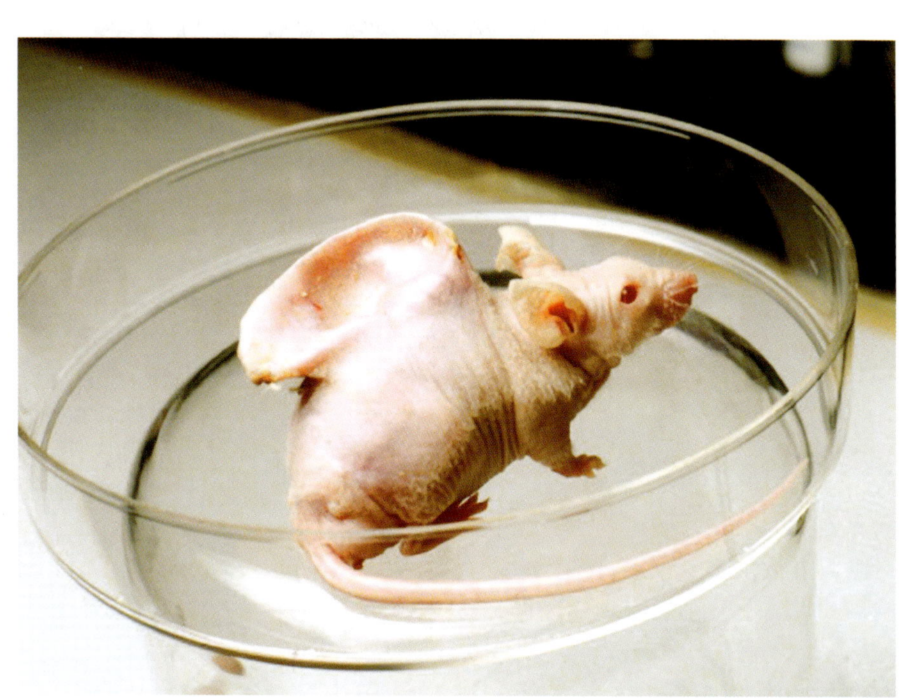

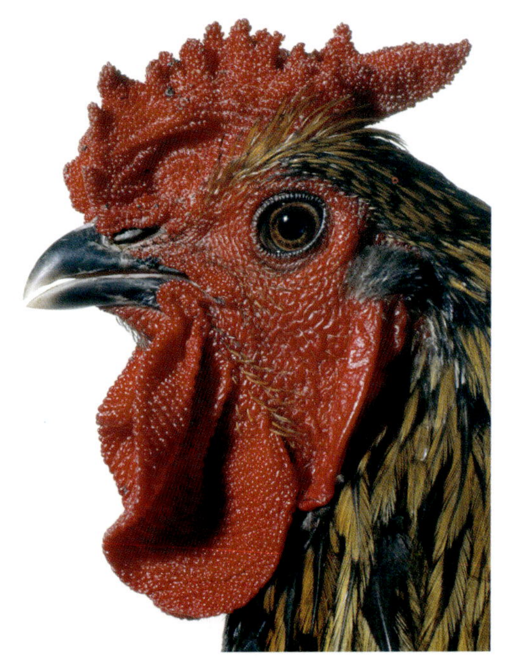

# A FREE SPIRIT TAKES LIBERTIES EVEN WITH LIBERTY ITSELF.

- FRANCIS PICABIA -

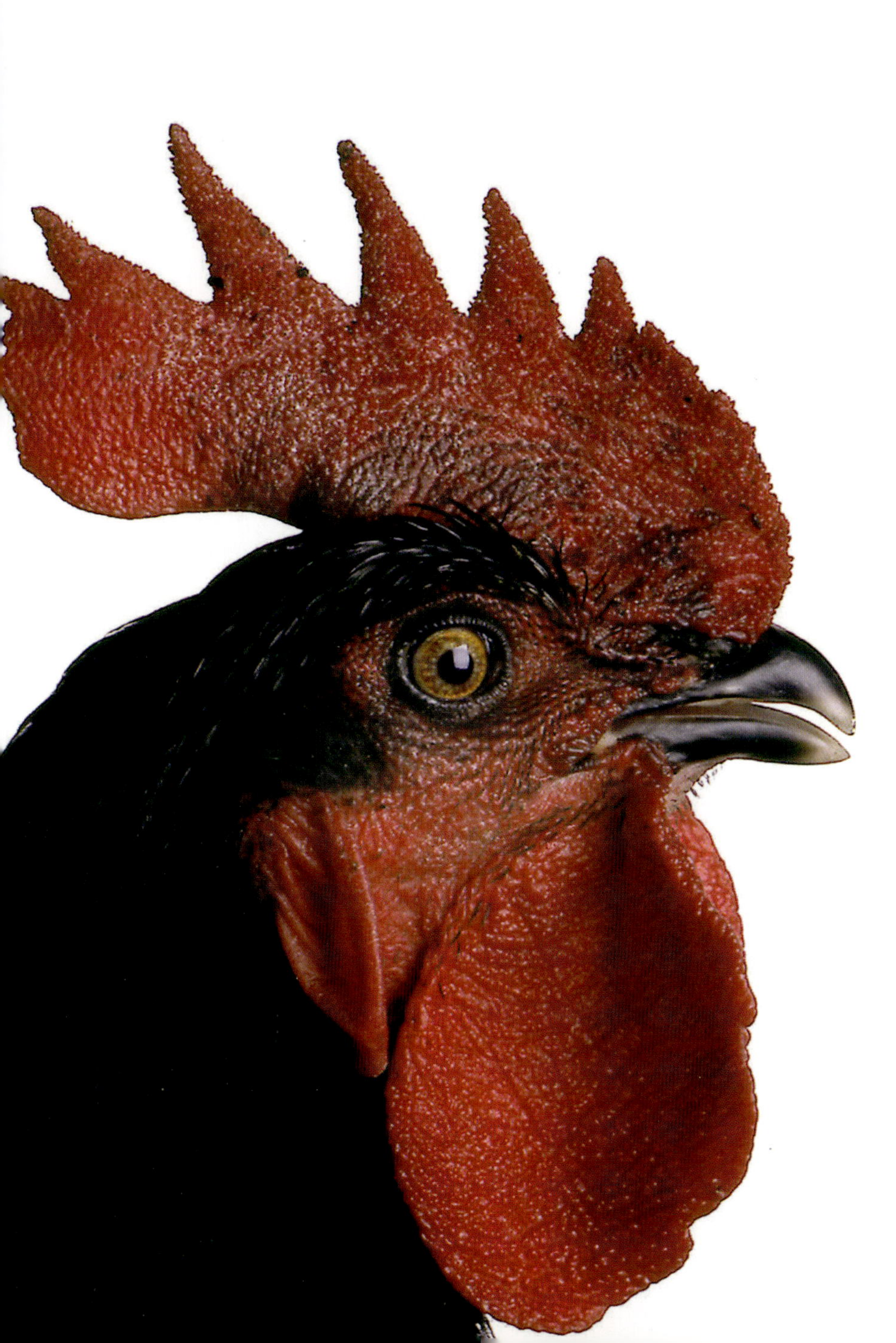

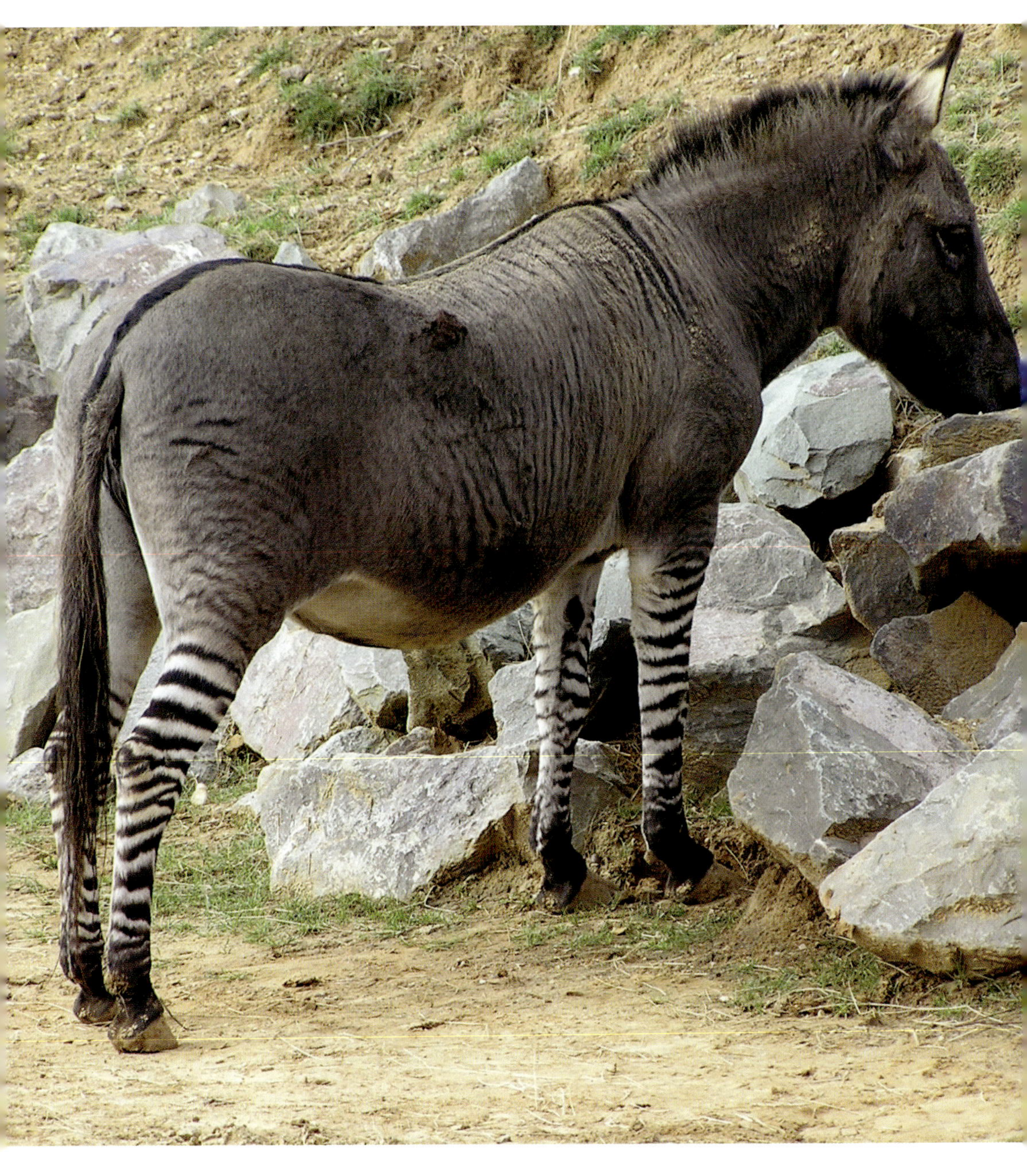

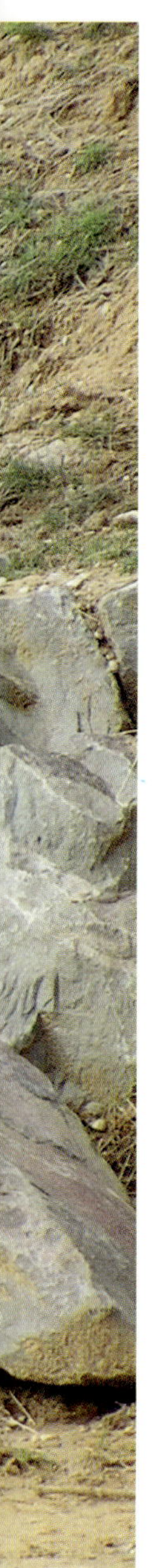

**ANTIPHOLUS OF SYRACUSE:**
**THOU HAST THINE OWN FORM.**

**DROMIO OF SYRACUSE:**
**NO, I AM AN APE.**

**LUCIANA:**
**IF THOU ART
CHANGED TO AUGHT,
'TIS TO AN ASS.**

- WILLIAM SHAKESPEARE -

BUT THERE BEING
NO SELF AND NO BARS THEREFORE
THE ZOO OF YOUR DEAR
FATHER HATH NO LION
YOU SAID YOUR MOTHER WAS MAD
DON'T EXPECT ME TO PRODUCE
THE MONSTER FOR YOUR BRIDEGROOM.

- ALLEN GINSBERG -

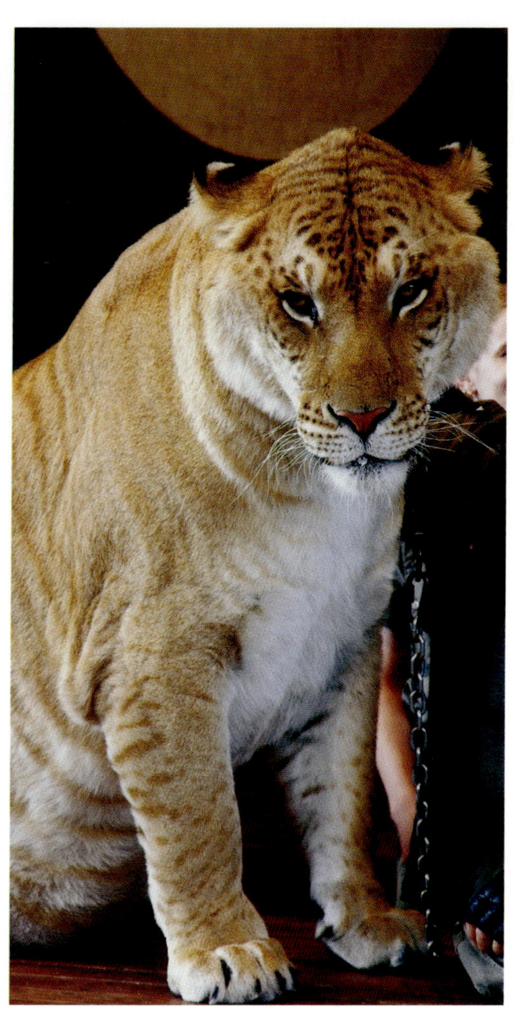
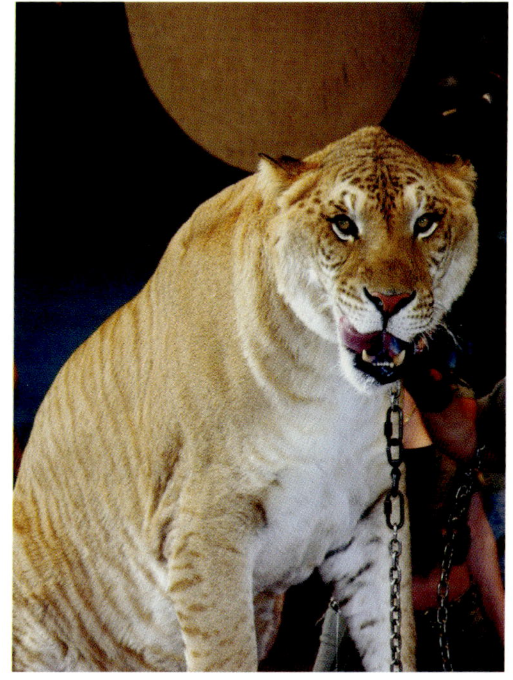

I AM NOT RAVING.
I AM NOT MAD.
I AM TELLING YOU THAT
MICROBES
HAVE BEEN RE-INVENTED IN
ORDER TO IMPOSE A NEW
IDEA OF GOD.

- ANTONIN ARTAUD -

WHY
SHOULD THE LAMP OR THE HOUSE
BE AN ART OBJECT,
BUT NOT OUR LIFE?

- MICHEL FOUCAULT -

If freedom involves driving the imaginary to the limits, one also needs to sign a pact with the devil as did Faust, the first technologist who formed an alliance with Mephisto, one of the devil's most elegant advocates. Faust's pact enabled him to work in the devil's laboratory. Which means, seeing the worst, he began charting the course beyond good and evil.

* * *

That work on hybridity runs the high risk of fabricating viruses susceptible to the creation of uncontrollable mutations and disasters. Wouldn't such a threat constitute true mutation? One needs to make welcome the greatest catastrophe if one wants to remain in that logic, to say "yes" like Nietzsche and to have the courage to see as he did where the radical evil is situated. Without wanting to push reflections on their merits and pitfalls, I must wonder why those questions, as far as I know, have never really been posed.

* * *

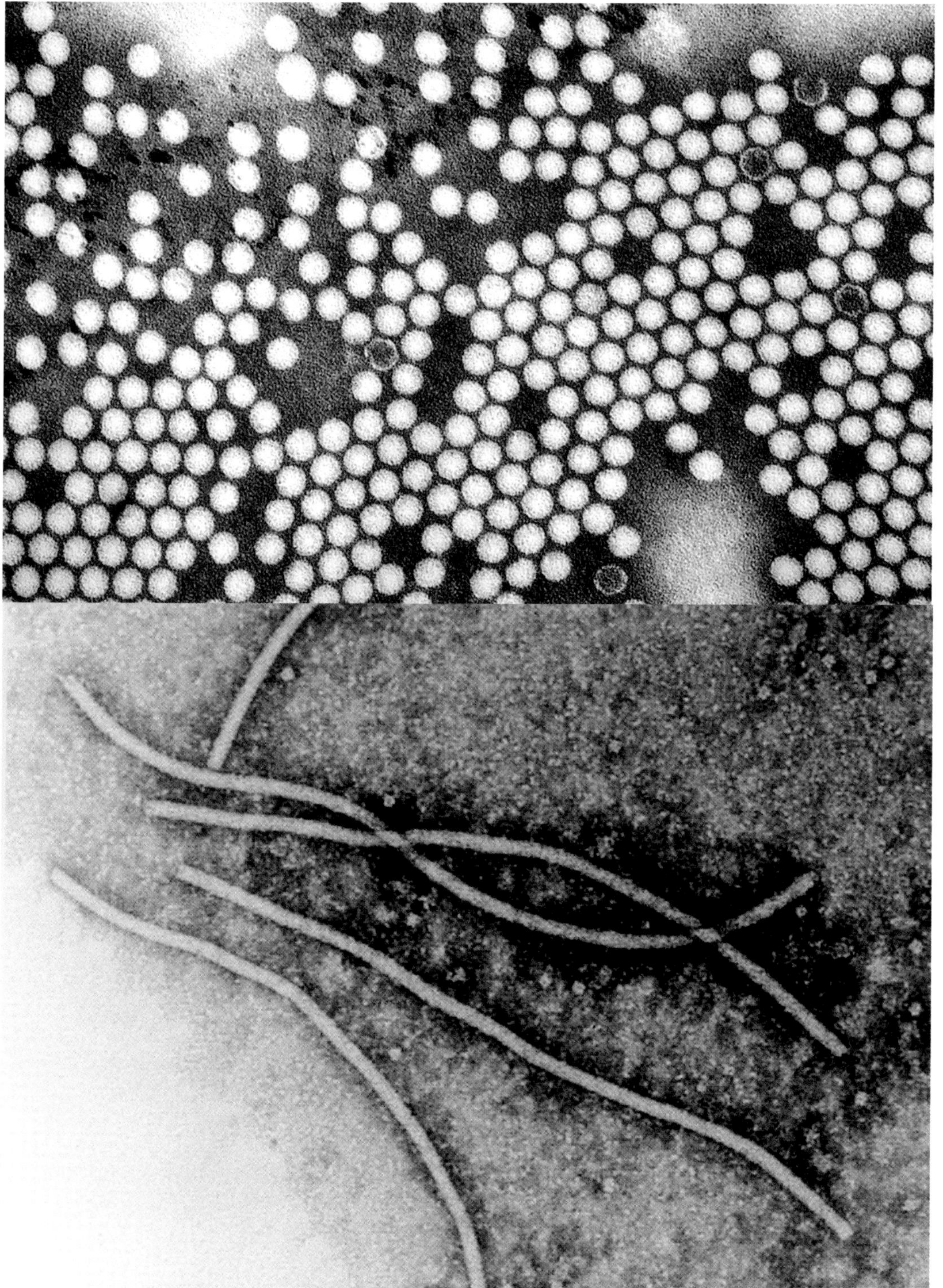

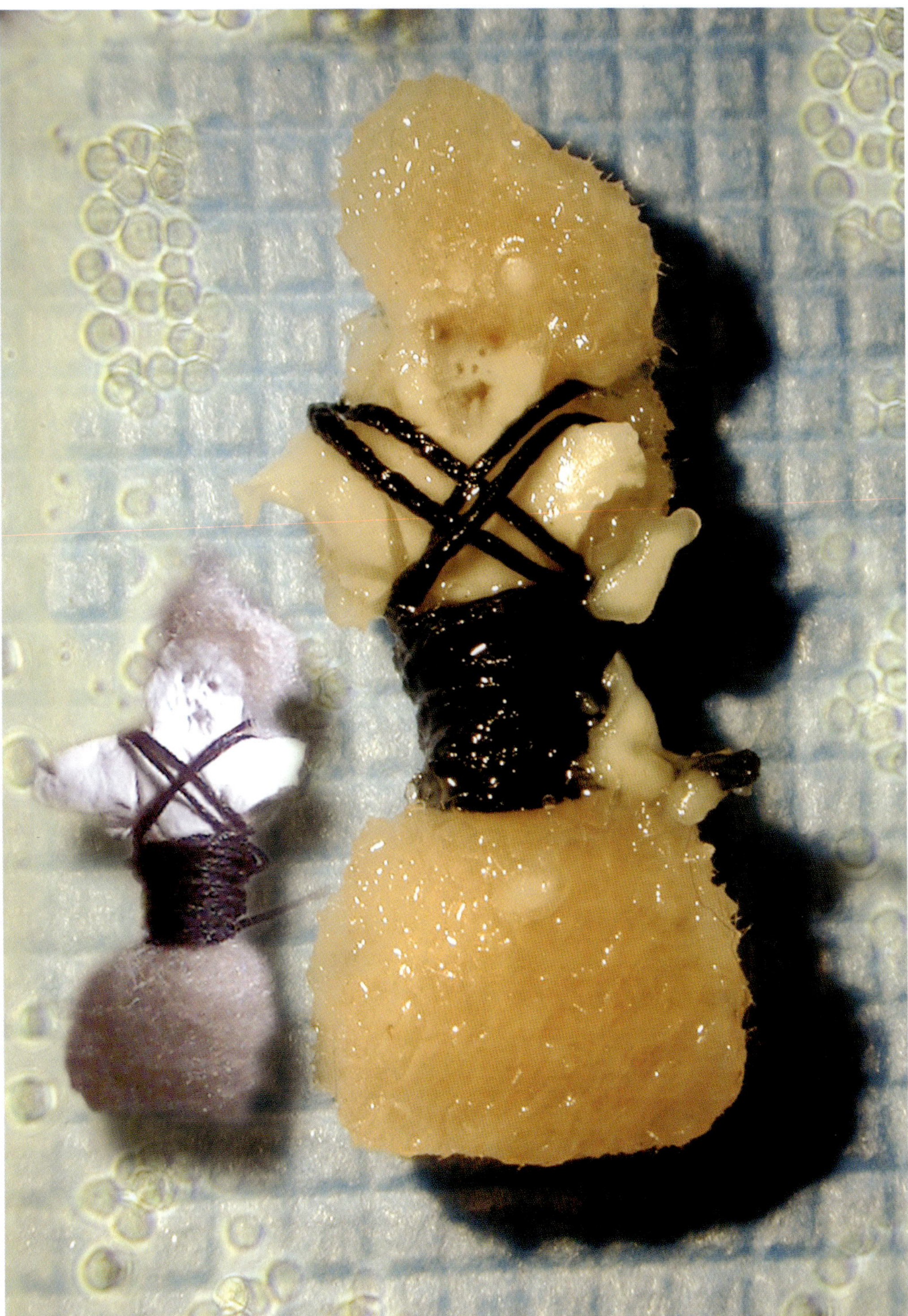

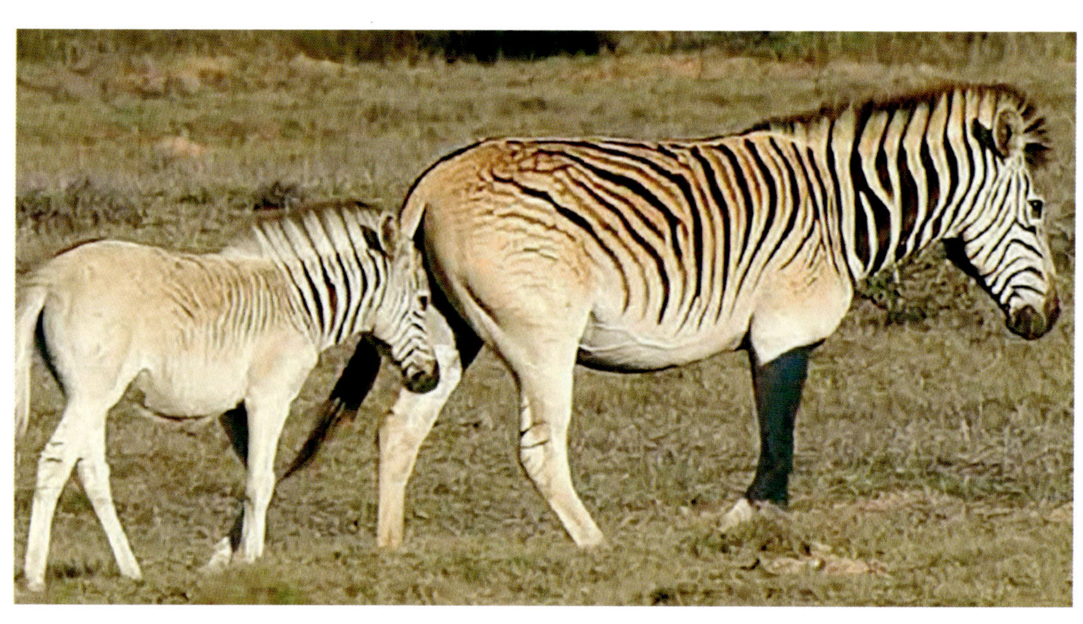

# IT IS NOT TRUE THAT THE *UNCONSCIOUS GOAL* IN THE EVOLUTION OF EVERY CONSCIOUS BEING (ANIMAL, MAN, MANKIND, ETC) IS ITS "HIGHEST HAPPINESS";

THE CASE, ON THE CONTRARY, IS THAT EVERY STAGE OF EVOLUTION POSSESSES A SPECIAL AND INCOMPARABLE HAPPINESS NEITHER HIGHER NOR LOWER BUT SIMPLY ITS OWN.

# EVOLUTION DOES NOT HAVE HAPPINESS IN VIEW, BUT EVOLUTION AND NOTHING ELSE...

- NIETZSCHE -

# A CAT IS NECESSARY TO COMPLETE THE HOME.

IT IS THE CAT THAT POLISHES THE FURNITURE, SOFTENS THE EDGES, GIVES THE PLACE AN AIR OF MYSTERY.

## IT IS THE ULTIMATE BIBELOT, THE SUPREME TOUCH.

- MALLARMÉ -

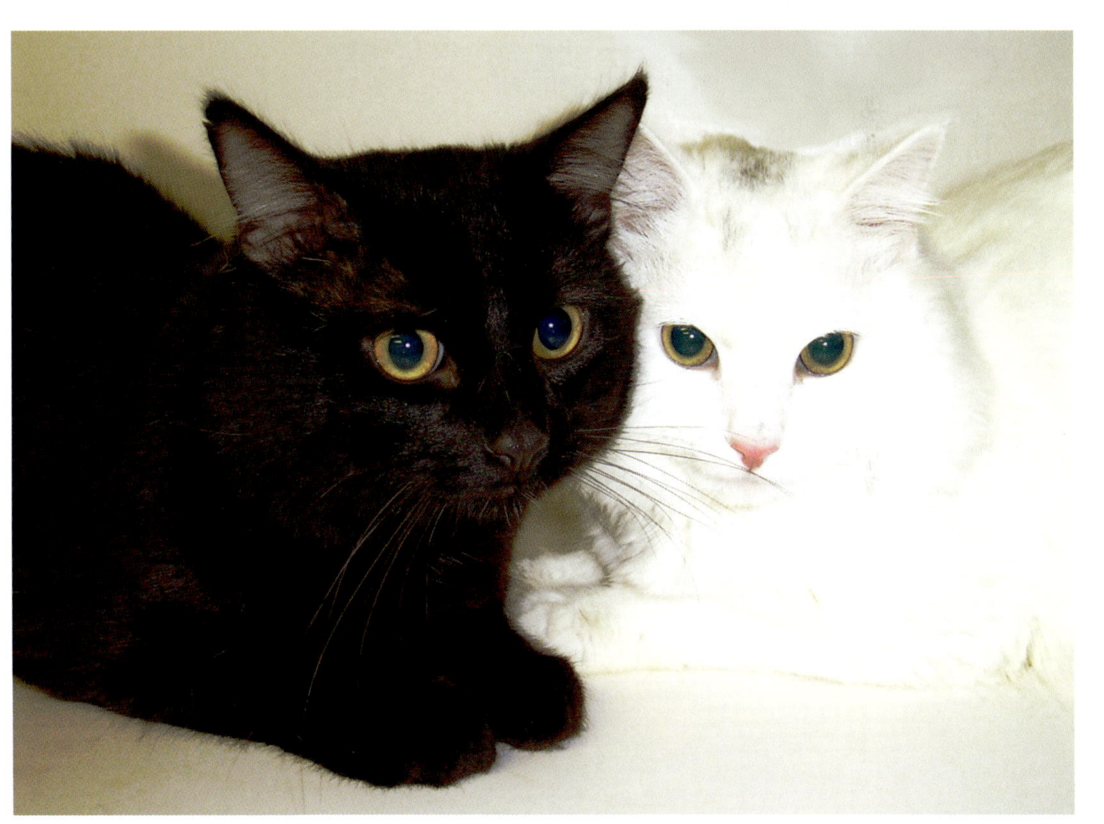

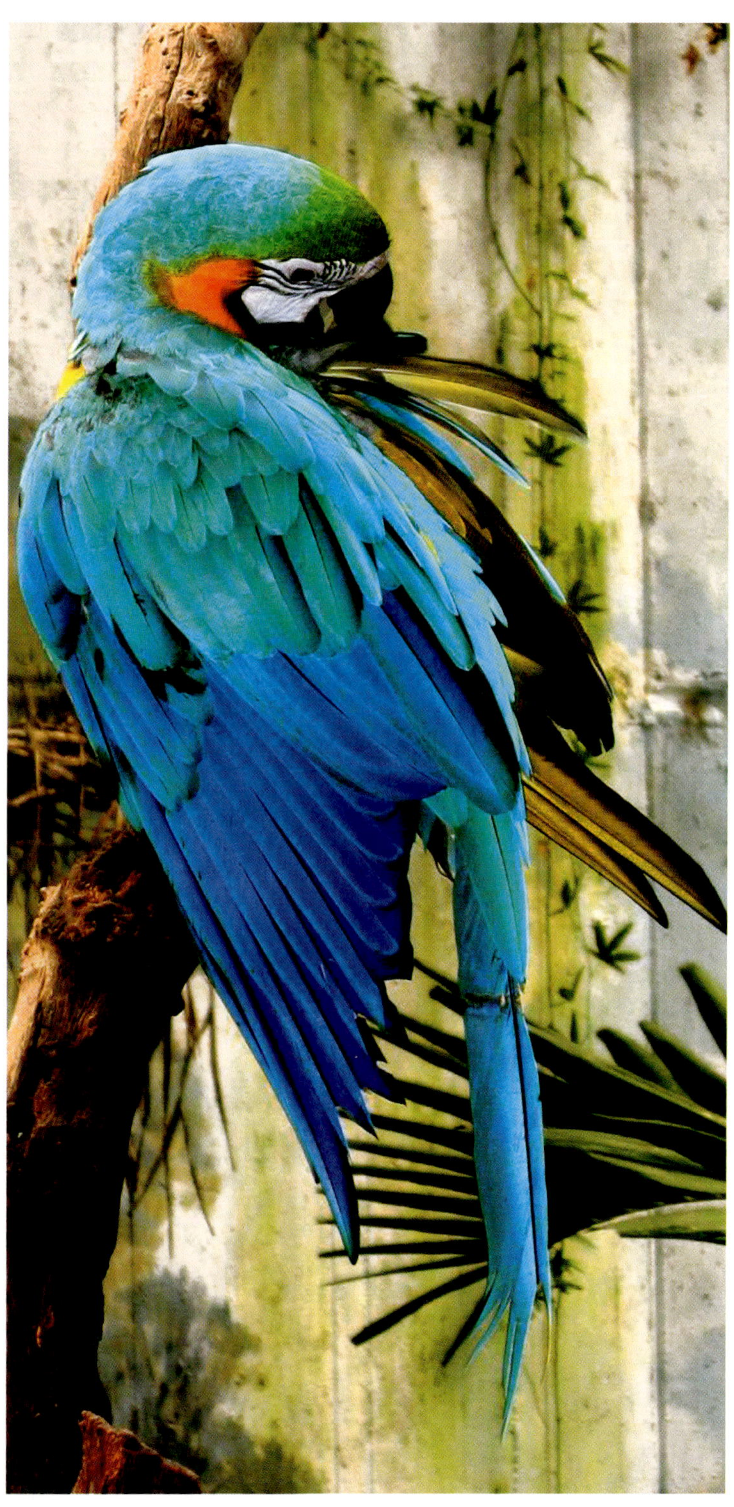

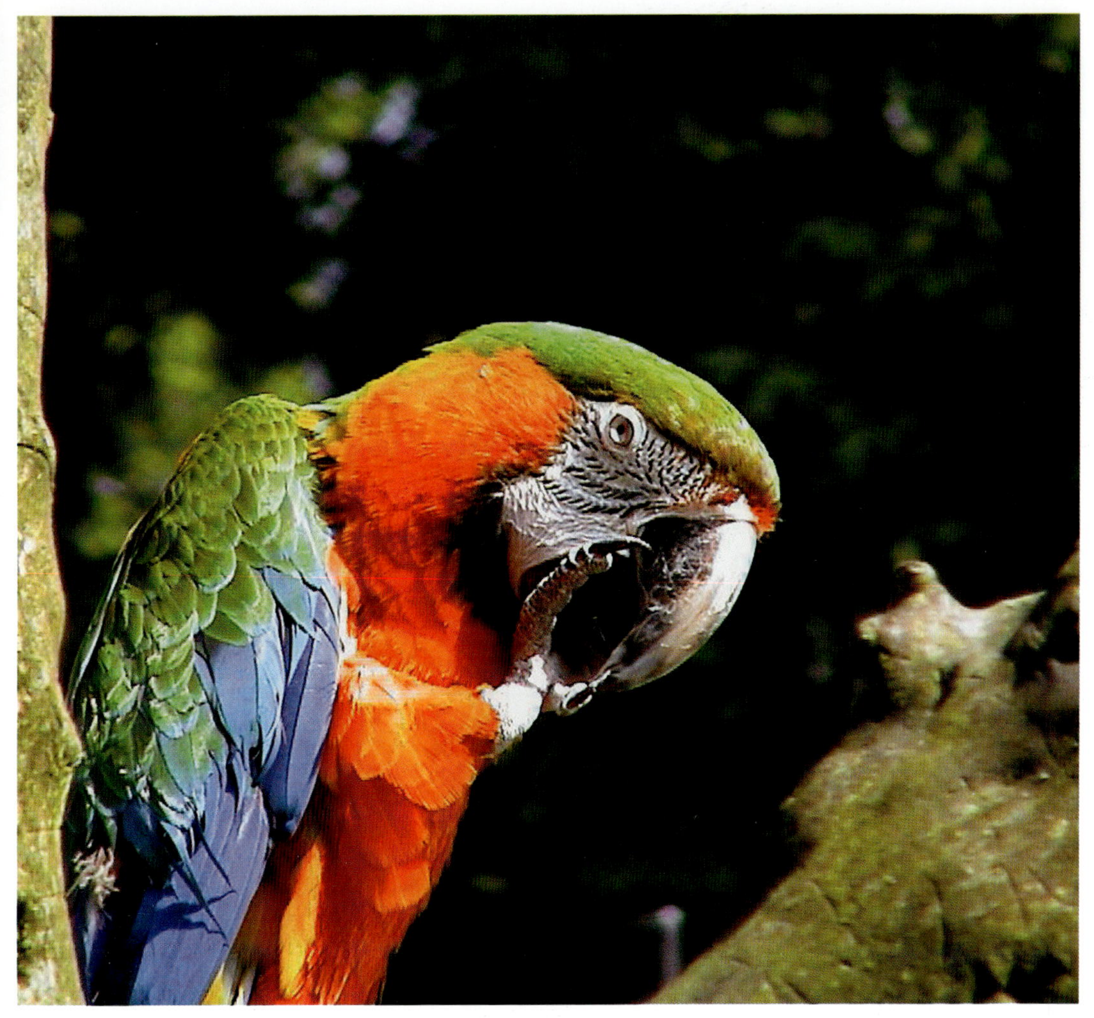

# PARROTS
# ARE WINGED MONKEYS.

- GUSTAVE FLAUBERT -

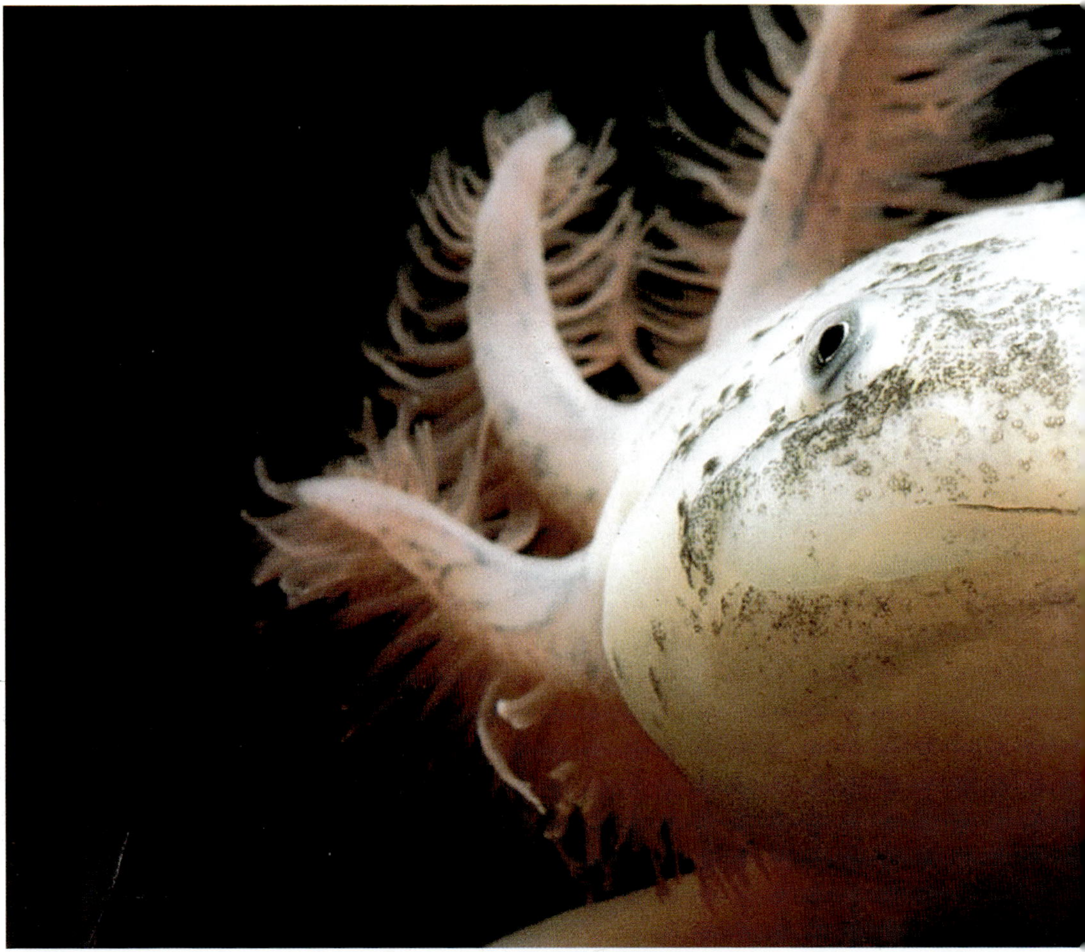

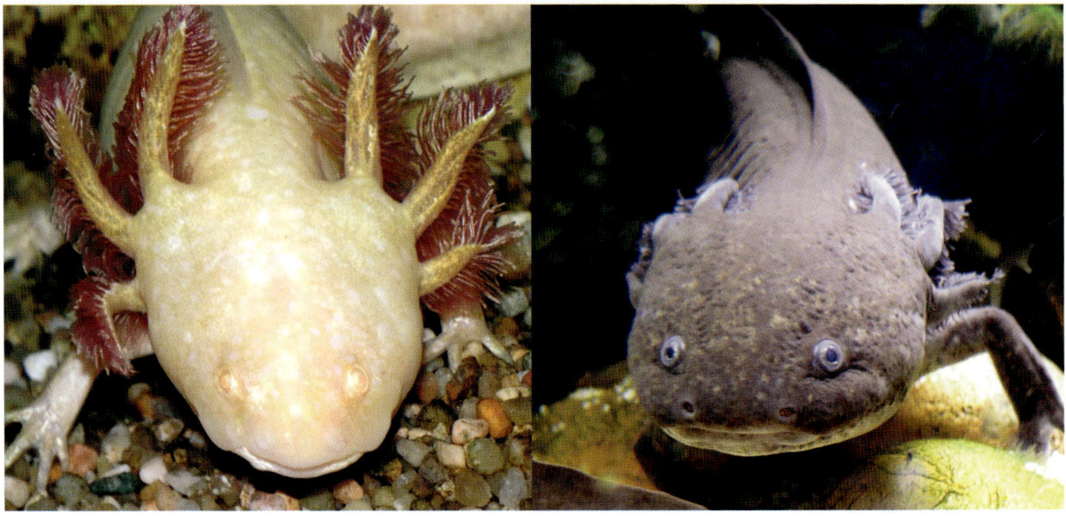

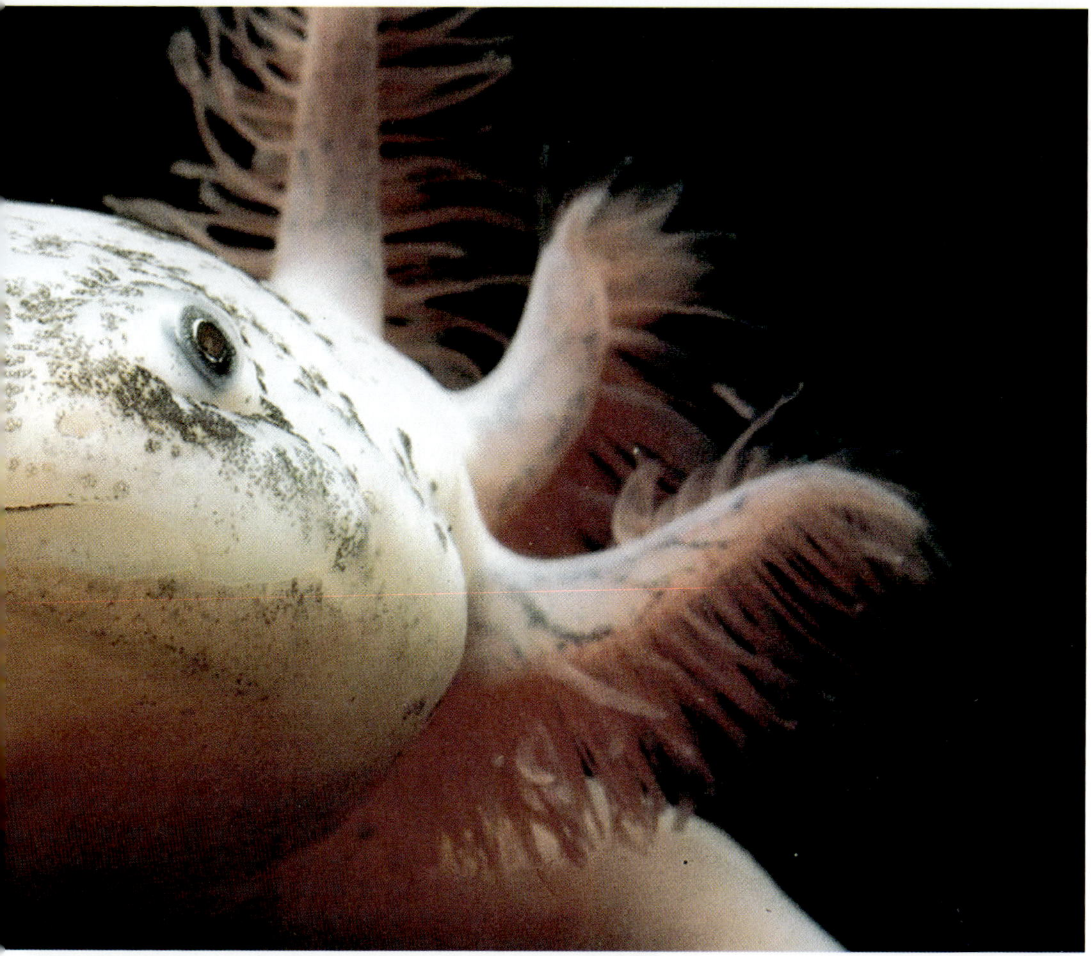

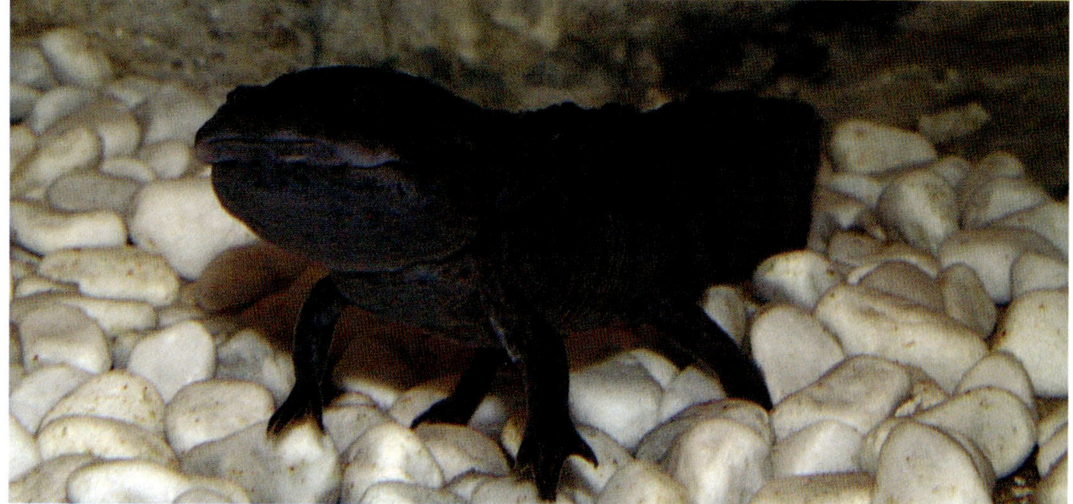

# I'M AFRAID THAT IF YOU LOOK AT A THING LONG ENOUGH, IT LOSES ALL OF ITS MEANING.

- ANDY WARHOL -

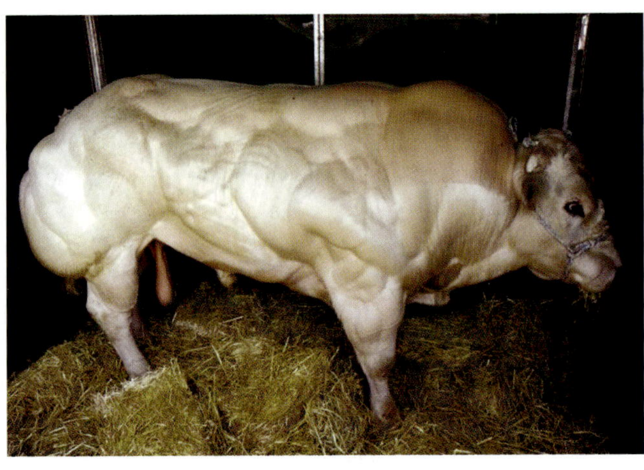

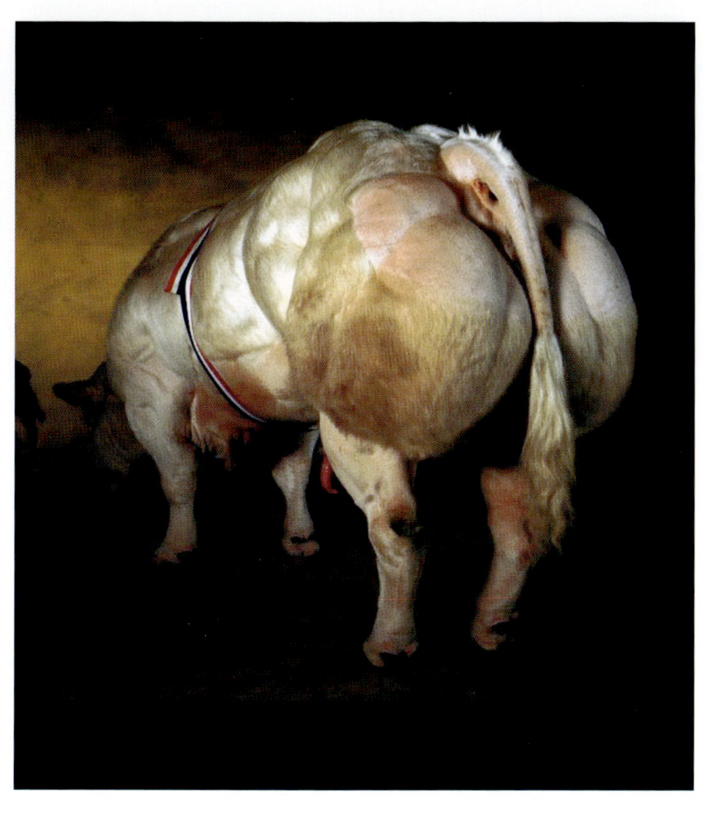

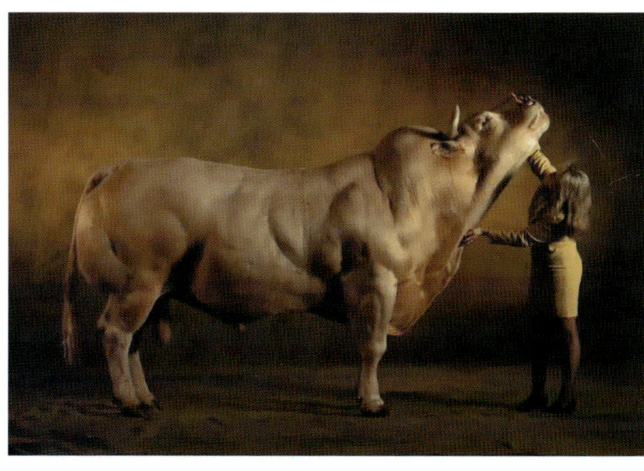

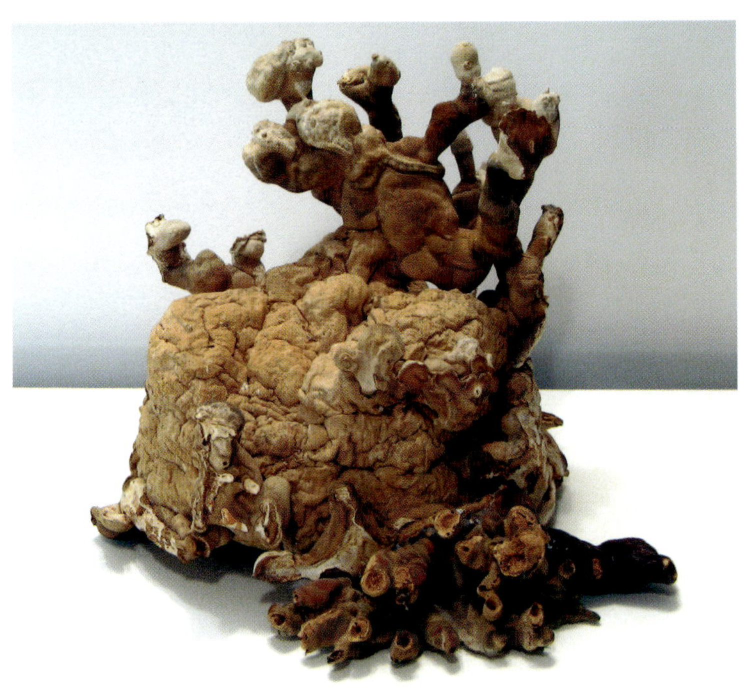

# NATURE EVOLVES INGENIOUS FORMS, OFTEN TECHNOLOGICALLY USEFUL.

EVERY BUSH, EVERY TREE,
CAN INSTRUCT US IN AND REVEAL NEW USES,
POTENTIAL APPARATUS,
AND TECHNOLOGICAL INVENTIONS
WITHOUT NUMBER.

- MOHOLY-NAGY -

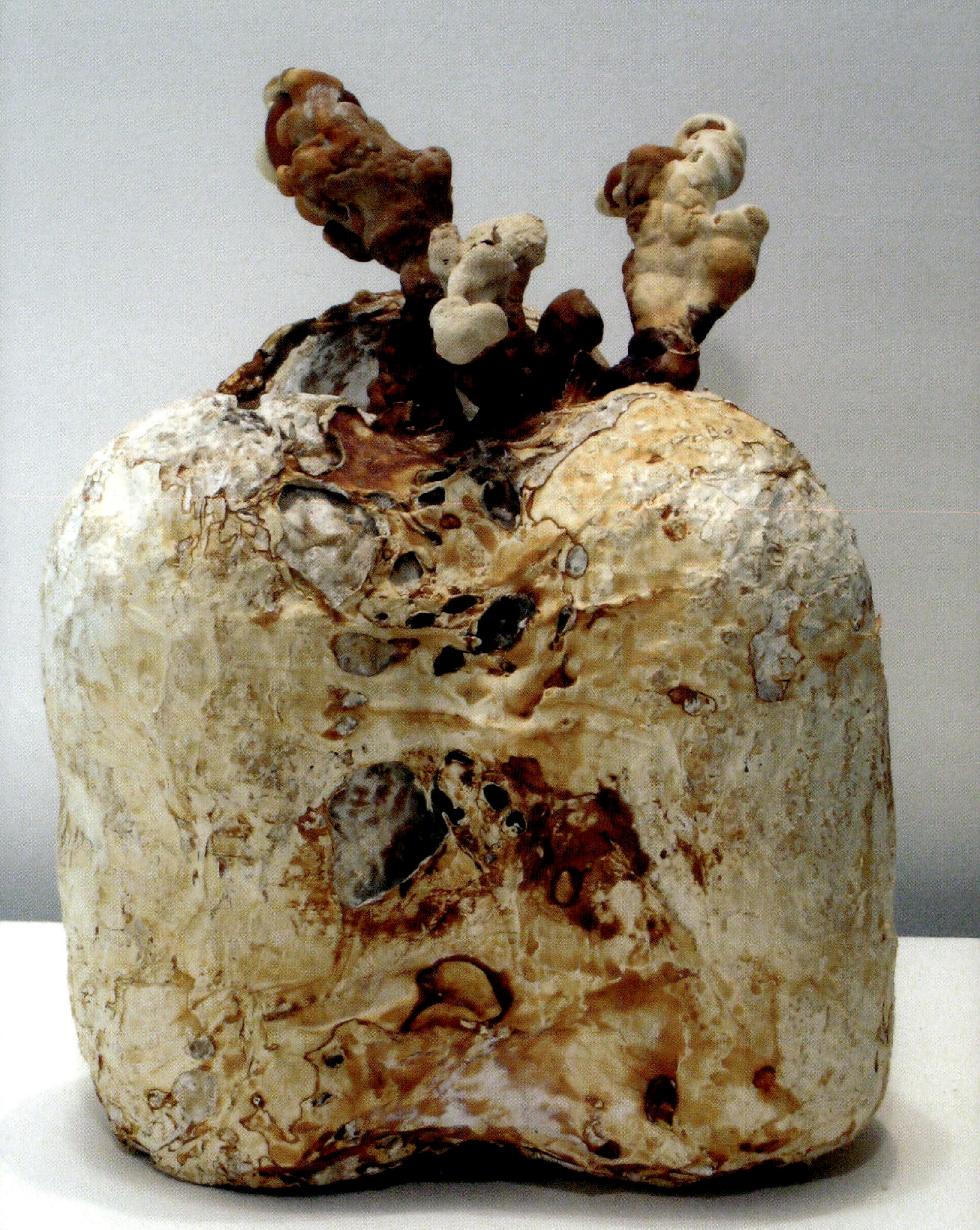

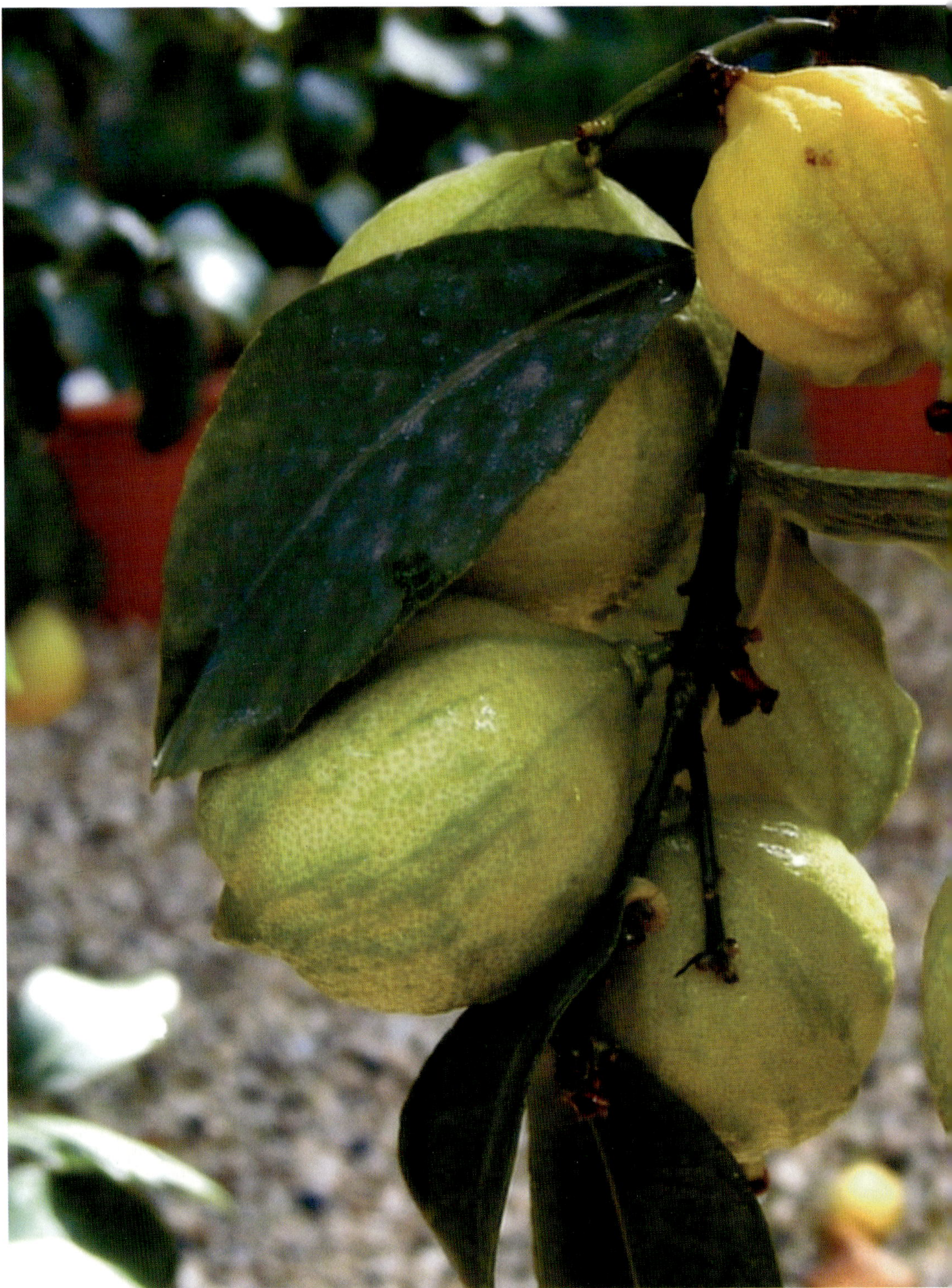

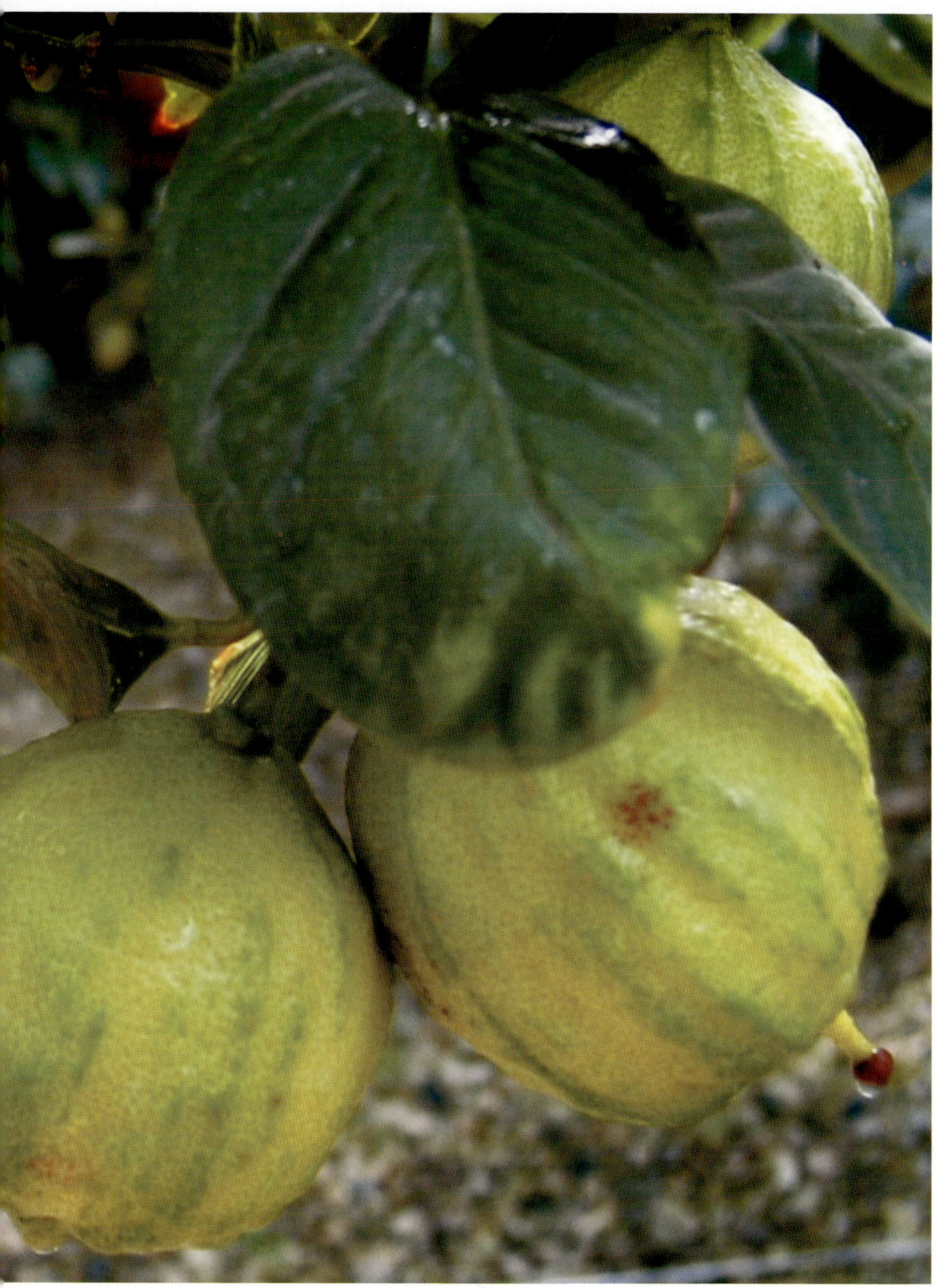

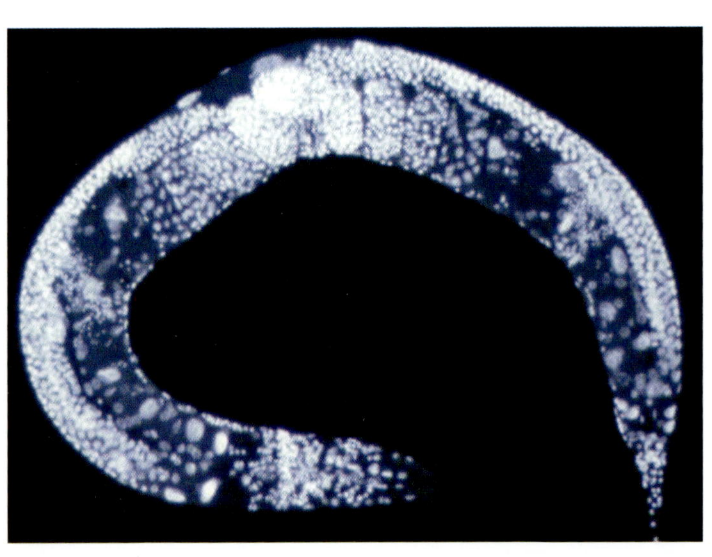

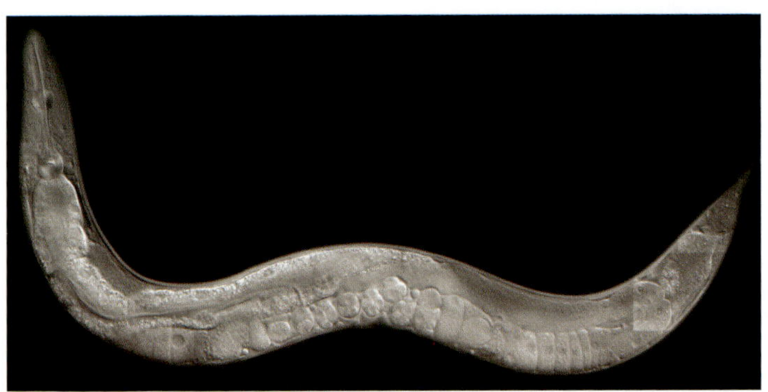

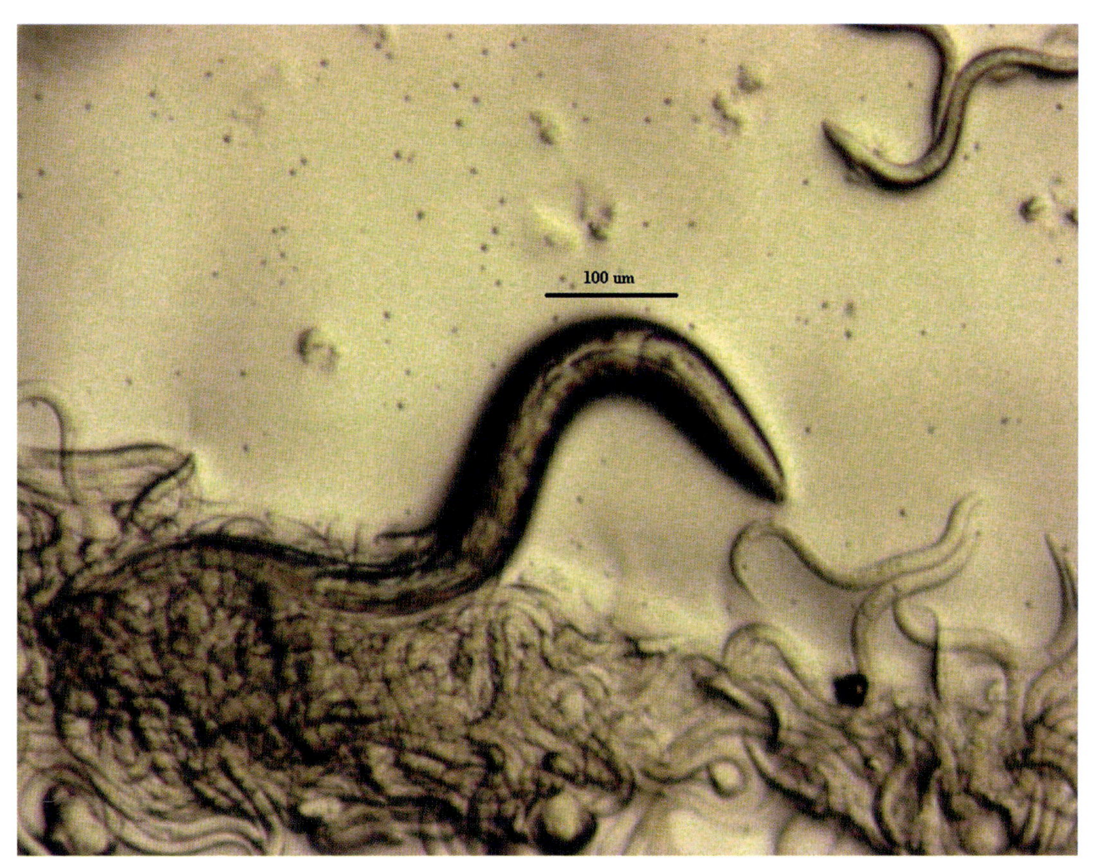

**EVERY ANIMAL CARRIES ITS ENVIRONMENT
ABOUT WITH IT LIKE AN IMPENETRABLE SHELL,
ALL THE DAYS OF ITS LIFE.**

- JAKOB VON UEXKÜLL -

To pretend that an animal has, like the human, a biography, forces us to wonder what this is. Biography is already a hybridity. Do the graphy and the bios go together, one on the side of death, the other, of life? One could perhaps say "thanatography". Derrida used to say "allothanatography", meaning that so-called autobiography is always also the history of the other, or of oneself as other. I remember that once Derrida was about to kill an insect and I said, "don't kill it, just make it go away". I try not to kill anything. I said, as English permits, "he's a person, too," crossing some spectral line.

* * *

Can an animal die? According to Heidegger it's not possible, for an animal can merely croak—an animal, like some food substances, perishes. Can one have a biography if one is not a being-destined-to-die? The notion of biography presupposes a whole metaphysics of the self and, if one shares it here with post-human or pre-human beings, or even post-animal beings, it is already an affair of breaking off from the historical dimension of self. Even for those who think they have a history, it is not obvious to share a biography with the new generation of post-creaturely beings, to find oneself there, at their side, us, the monsters that we are.

* * *

THE POET,
LIFTED UP WITH THE
VIGOR OF HIS OWN
INVENTION,
DOES GROW IN
EFFECT ANOTHER
NATURE,
IN MAKING THINGS EITHER
BETTER THAN NATURE
BRINGS FORTH, OR,
QUITE ANEW,
FORMS SUCH AS
NEVER WERE IN
NATURE.

- PHILIP SIDNEY -

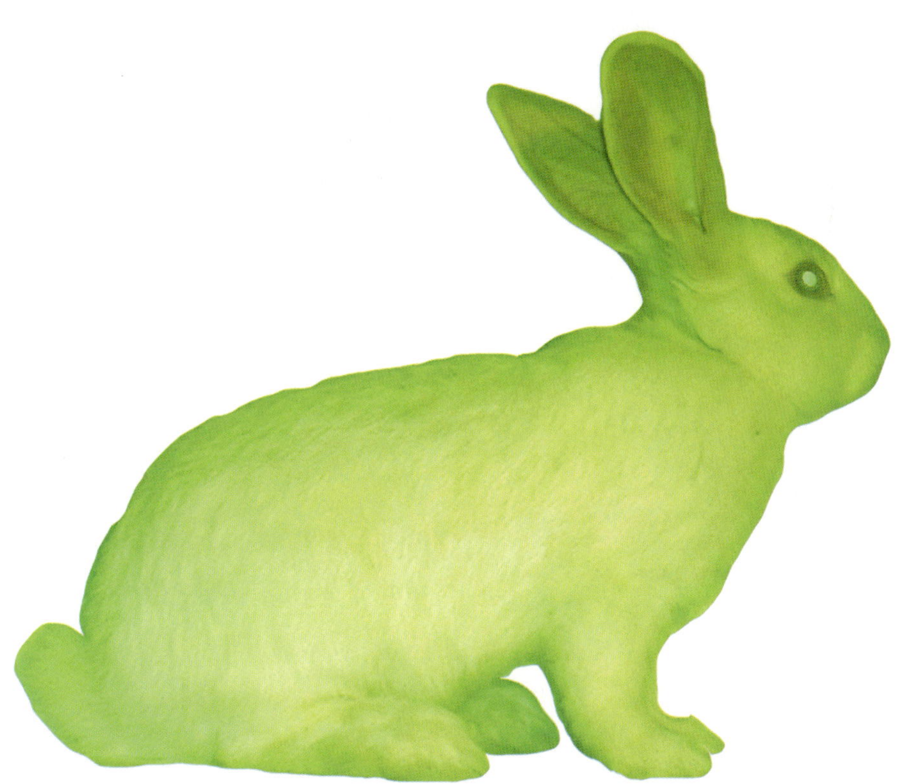

# PHOTOGRAPHS

[11] **Cubic Watermellon** © DR.

[14] **Nude mouse.** © DR.

[17] **Featherless chicken**. Courtesy Avigdor Cahaner

[20] **Cloned Dog**. Courtesy Byeong-Chun Lee.

[24] **Glofish.** © DR.

[27] **Geep.** Courtesy Nature magazine.

[28/29] **Frizzled Chicken**. Courtesy Klaas van der hoeck, Luuck Hans, Eduardo Kac, André Coupet.

[31] **Fatherless Mouse.** ©DR.

[34] **Steichen Strain Delphinium.** Photo Edward Steichen ©Johanna Steichen

[38/39] **Wolphin.** ©DR.

[42/43] **Dolly, the cloned sheep.** ©DR.

[45] **Quail-chicken Chimera**. Courtesy CNRS photothèque.

[48/49] **Reconstructed aurochs.** © DR.

[50/51] **Butterflies with modified wing patterns**. Courtesy Marta de Menezes.

[52/53] **Sphynx Cat**. © DR.

[56/57] **Longhorn clone**. © DR.

[58/59] **Hybrid Irises**. Courtesy George Gessert

[63] **Transgenic monkey.** © DR

[69/70] **White Tigers.** © DR.

[66/67] **Cactus sculptures**. Courtesy Philippe Obliger.

[68/69] **Savannah cat.** © DR.

[70/71] **Antennapedia flies**. © DR.

[74] **Blue rose.** Courtesy Steve Chandler (Florigene)

[79] **Mouse with Human Ear**

[80/81] **Cosmopolitan Chicken.** Courtesy Koen Vanmechelen.

[82] **Zeedonk.** Courtesy Colchester Zoo, UK.

[85] **Liger.** © DR.

[87] **Synthetic virus.** © DR.

[88] **Semi Living Worry Doll.** Courtesy Oron Catts, Ionat Zurr .

[92] **Quagga.** Courtesy Quagga Project Association.

[95] **Hypoallergenic cat.** © DR.

[96/97] **Catalina Macaws.** © DR.

[98/99] **Laufberger's Axolotl.** © DR.

[100/101] **Schwarzenegger Cow.** © DR.

[102/103] **Mushroom Sculpture.** Courtesy Philip Ross

[104/105] **Bizzarria Orange.** Courtesy Petr Broza.

[106/107] **Methuselah Worm.** © DR.

[109] **Alba, the Bunny**. Courtesy Eduardo Kac.

[11] **Gottfried Wilhelm Leibniz** - "A Résumé of Metaphysics" (c. 1697), in: *Leibniz: Philosophical Writings*, G. Parkinson, ed. (London an Melbourne: Dent, 1973), p. 145.

[14] **Lucian of Samosata** - *Trips to the Moon / The True History)*

[17] **Diogenes Laertius** - *Lives of Philosophers* 6.40

[21] **Gertrude Stein** - *The Geographical History of America or the Relation of Human Nature to the Human Mind* [1936],

[25] **Paul Eluard** - *Capital of Pain* , 1926

[26] **Ovide** - *Metamorphosis*, Book XV

[28] **Rimbaud** - *Letter to Demeny* dated 17 April 1871

[30] **Sapho** - *Love Songs Of Sappho,* Translated by Paul Roche, Prometheus (1998)

[35] **Carl Sandburg** - "Complete Poems" (New York: Harcourt, Brace & World, 1950) p. 673. (*Written at Steichen's Umpawaug Plant Breeding Farm, in Connecticut, 1933*).

[38] **Martin Buber** - *I and Ihou* (1923), trans. Ronald G. Smith (2d ed.; New York: Charles Scribner's Sons, I 958), p. 97.

[44] **Alfred Hitchcock.** - *Transcript of Alfred Hitchcock's speech at National Press Club in Washington DC on 18 March 1963*

[49] **Elizabeth Bishop** - *Geography III, One Art.*

[50] **Flávio de Carvalho** - *Experiência N. 2* (São Paulo: Irmãos Ferraz, 1931), p. 115.

[53] **Lowell Edmunds** - *The Sphinx in the Oedipus Legend.* Königstein im Taunus: Hain, 1981.

[57] **Paulo Leminski** - *Caprichos e Relaxos* (São Paulo: Brasiliense, 1983), p. 98. (1944-1989)

[58] **George Gessert** - "Breeding for Wildness", in: Oron Catts, editor. *The Aesthetics of Care,* proceedings of the homonymous Symposium realized on the occasion of the Biennale of Electronic Arts Perth (BEAP) 2002 and published by SymbioticA, Perth, Australia, p. 31.

[62] **Mary Shelley** - *Frankenstein*

[65] **e.e. Cummings** - *73 Poems #67 (1963)*

[66] **Vicente Huidobro** - *Square Horizon*, 1917

[69] **Montaigne** - *Essays*, Book 2

[73] **Aristotle** - *Parts of Animals,* V

[75] **Novalis** - *Heinrich von Osterdingen;* 1876.

[78] **NamJune Paik** - « Art & Satellite », in : Dorine Mignot, ed., *The Luminous Image* (exhibition, catalogue), Stedelijk Museum, Amsterdam, 1984, p.68

[80] **Francis Picabia** - quotes in *Picabia, 1879-1953, Kunsthalle Bern, 1962.*

[82] **William Shakespeare** - *Comedy of Errors ,* Act 2, Scene 2.

[85] **Allen Ginsberg** - *The Lion for Real,* in: *Collected Poems, 1947-1980* (New York: Harper, 1985)

[86] **Antonin Artaud** - *To have done with the judgement of God,* 1947.

[90] **Michel Foucault,** *On the Genealogy of Ethics: An Overview of Work in Progress*, in: Hubert L. Dreyfus and Paul Rabinow, eds., "Michel Foucault: Beyond Structuralism and Hermeneutics" (Chicago: U Chicago, 1982). p. 236.

[92] **Friedrich Nietzsche** - *Daybreak: Thoughts on the Prejudices of Morality*, Book II, N. 108; Maudemarie Clark, Brian Leiter, eds.(Cambridge: Cambridge University Press, 1997), p. 63.

[94] **Stéphane Mallarmé** - quoted in: Docquois, Georges: *Bêtes et Gens de Lettres,* 1895 (Paris, Flammarion,).

[99] **Gustave Flaubert** - quotes by Marie-Jeanne Durryin in *Flaubert et ses Projets Inédits M.S.* (Bibliothèque Nationale, folio 388 (Paris; Nizet, 1950).

[100] **Andy Warhol** - in *Andy Warhol.* Rétrospective, Centre Georges Pompidou, 1990, pp. 457-467

[102] **László Moholy-Nagy** - *Why Bauhaus Education?* Shelter (March 1938): 8-21, quote p. 15, 12.

[106] **Jakob von Uexküll** - quoted in Ernst Cassirer, *The Problem of Knowledge,* p. 202.

[110] **Philip Sidney** - *Defense of Poesy,* c. 1583

### Cubic Watermelon

In 2001 farmers in the southern Japanese town of Zentsuji started to grow cubic watermelons. Lack of space is a known problem in Japan. While a round and large fruit can take up a lot of room in a refrigera*tor, the geometric fruit perfectly fits the Japanese appliance*s. This luxury food is not eaten by regular customers but is instead given as a gift to business associates, that is, people to whom one has an obligation because of a favor they have done. The cubic watermelon may cost up to ten times the price of a conventional watermelon.

### Nude mouse

The first nude mouse appeared at the Virus Laboratory, Ruchill Hospital, Glasgow in 1962, in Dr. Grist's laboratory in a closed but not deliberately inbred albino stock. The nude mouse was brought to Denmark by Dr. Rygaard in 1968, and in 1969, breeding pairs were obtained. The line was backcrossed 3-4 generations. Nude mice have since been bred continuously in laboratories worldwide. A nude mouse is a genetic mutant that has a deteriorated or removed thymus gland, resulting in an inhibited immune system. The phenotype, or main outward appearance of the mouse is a lack of body hair, which gives it the "nude" nickname. The nude mouse is widely used in research because it can receive many different types of tissue and tumor grafts, as it mounts no rejection response.

### Featherless chicken

In 2002 Israeli scientist Avigdor Cahaner developed a new featherless broiler chicken. Using several cycles of cross-breeding, he introduced the natural mutation "scaleless", which was found in 1954 in a slow-growing small-body breed of chickens, into a breed of fast-growing large-body chickens (broilers) bred for meat production. This resulted in completely naked broilers. The goal of this research is to increase both tolerance to heat stress and industrial processing efficiency. The right-page photograph shows a small-body female from the original population of mutants (right), and a large-body female from the line he developed (left). Both are at the same age.

### Cloned Dog

In 2005 scientists at Seoul National University cloned dogs by somatic-cell nuclear transfer. While several mammals — including sheep, mice, cows, goats, pigs, rabbits, cats, a mule, a horse and a litter of three rats — had already been cloned at that point by transfer of a nucleus from a somatic cell into an egg cell (oocyte) that had its nucleus removed, this technology had not been successful in dogs because of the difficulty of maturing canine oocytes *in vitro*. The Korean scientists cloned two Afghan hounds by nuclear transfer from adult skin cells into oocytes that had matured in vivo. The cloning technique used is not efficient. It took nearly 2,000 eggs to make some 1,000 embryos — all of which produced just one healthy puppy. The photo shows Snuppy, the first cloned dog, at 67 days after birth (right), with the three-year-old male Afghan hound (left) whose somatic skin cells were used to clone him. Snuppy's surrogate mother was a yellow Labrador. Snuppy is genetically identical to the donor Afghan hound.

### GloFish

GloFish is the trademarked name of a transgenic version of the Zebra Danio, or Zebra Fish, the aquatic equivalent of a white lab mouse widely used in research. Originally created and patented by Gong Zhiyuan and his group at the National University of Singapore, the transgenic Zebra Fish had its exclusive rights for international marketing purchased by the American firm Yorktown Technologies, who launched the ornamental fish in the United States in 2003. While the regular zebra fish is mostly flat black and silver, the GloFish expresses a red fluorescent protein from a sea anemone and looks red under regular light. After the initial launch of the red GloFish, the company also introduced green and orange varieties. Initially, some expressed concerns about the open sale of genetically modified animals. In the United States, it is the Food and Drug Administration that asserts jurisdiction over genetically modified animals. The FDA determined that "In the absence of any clear risk to the public health, the FDA finds no reason to regulate these particular fish." This position was contested and a lawsuit filed in 2004 by the Center for Technology Assessment and the Center for Food Safety. Meanwhile, commercialization of the tropical fish went forward in the United States and in several countries such as Taiwan, Malaysia, and Hong Kong. While the regular Zebra Fish sells for less than $1, each GloFish sells for approximately $5.00 and is widely available at pet shops in the United States.

## Geep

Created in 1984 by Danish embryologist Steen Willadsen, the geep is a sheep-goat chimera, produced by combining the embryos of a goat and a sheep in a petri dish. With cells of both sheep and goat origin, it has the head of a goat and the woolly upper torso of a sheep. The Geep is not a sheep-goat hybrid, which sometimes results from the mating of a goat and a sheep. The geep is a mosaic of uneven goat and sheep parts: hairy where the animal is goat and woolly where it is sheep. The geep may be fertile, but the offspring of a geep will be either sheep or goat, because the geep's reproductive organs are formed from either sheep or goat embryo.

## Frizzled Chicken

Of unknown origin, the frizzled chicken has curling feathers and was developed as a race in England in the 19th century. The frizzy feature is a mutation in the chicken in which the feathers grow so that they curve outward, instead of lying smoothly along the bird's body. It can occur in many breeds. Charles Darwin classified them as Frizzled or Caffie Fowls. Frizzled chicken are bred principally for exhibitions, in which they must be bay (reddish-brown), black, buff (beige), blue, red or white in plumage color. To conform to race standards, the main points for exhibition purposes are the curl, which is most pronounced on feathers not too broad; the purity of color in plumage, and correctness in leg color; i.e., yellow legs for the white, red, or buff, and yellow or willow for other varieties.

## Fatherless Mouse

In 2004 a fatherless mouse was created by Japanese and Korean scientists without the use of sperm. The team, led by Tomohiro Kono of Tokyo University of Agriculture, made the animal, named Kaguya, by combining the nucleus of one female's egg with that of another, essentially creating a mouse with two mothers. Although plants, some fish and reptiles, frogs, insects and occasionally chickens can all procreate without a partner, the birth of Kaguya showed that a healthy mammal known as a parthenote can also be created without any male help. Kaguya was named after a Japanese folk tale, in which the moon-born princess Kaguya (Kaguya-hime) is found as a baby inside a bamboo stalk. After nearly 460 attempts at growing embryos, ten live pups were born and just one of those survived to adulthood. Kaguya had conventionally fathered babies of her own.

## Steichen Strain Delphinium

In its edition of July 6, 1936, Time Magazine stated: "What critics called 'the most amazing exhibit of delphinium ever seen in this country' was displayed at Manhattan's Museum of Modern Art by famed Photographer Edward Steichen, who is also president of the Delphinium Society of America. Florist Photographer Steichen explained that his blossoms were the result of 26 years' experiment and crossbreeding." Edward Steichen was the first modern artist to create new organisms both through traditional and artificial methods, to exhibit the organisms themselves in a museum, and to state that genetics is an art medium. To create his flowers (delphiniums) he hybridized them by hand but also employed chemicals to provoke mutations.

## Wolphin

In 1985 an Atlantic Bottlenose Dolphin gave birth at the Sea Life Park in Hawaii. The baby, who was given the name Kekaimalu, was not only darker than a dolphin but her snout was uncharacteristically short. When baby Kekaimalu opened its mouth another surprise awaited. Whereas an Atlantic Bottlenose Dolphin has 88 teeth, Kekaimalu displayed just 66 teeth. This was no ordinary dolphin. The dolphin had, unknown to the Park staff, engaged

in a torrid affair with one of her co-stars — who just happened to be an 18 year old, 2000 pound false killer whale. The result of this interspecies union was a new breed of sea mammal — half dolphin, half whale. It's true that herds of false killer whales and bottlenose dolphins interact in the wild and that the wolphin (also written "wholphin") was not intentionally bred by Park staff, but reports of natural wolphin hybrids are unsubstantiated. It was the fact that the bottlenose dolphin and the false killer whale were artificially kept together in a closed environment by humans at the Sea Life Park that led to the birth of the known wolphins. The so-called "false killer whale" is in reality a kind of dolphin so the wolphin is also a new a kind of dolphin. In 2004, Kekaimalu had a healthy daughter, called Kawili Kai. The father is a male bottlenose dolphin.

## Dolly, the cloned Sheep

While it is generally acknowledged that biologist J. B. S . Haldane coined the term "clone" in 1963, in a speech entitled "Biological Possibilities for the Human Species of the Next Ten-Thousand Years," it was on the 5th July 1996 that Dolly was born.  In 1997 she was announced to the world. For the first time in history, an adult mammal produced an offspring without an egg being fertilized by a sperm. Contrary to popular belief, Dolly was not born with the genetic characteristics of an old animal. In an attempt to show that she was physiologically normal, scientists bred her with a Welsh mountain ram, resulting in 6 lambs.   Their first, Bonny, was born in the spring of 1998. The next year they had twins. They had triplets the year after that. In 2001 it was found that Dolly had arthritis. In an unrelated development, it was found in 2003 that she had sheep pulmonary adenomatosis (SPA), an incurable disease caused by a virus that induces tumors to grow in the lungs of affected animals.  At that point scientists decided that putting her to sleep was best and ended her life. Subsequently, she was taxidermied and permanently displayed at the National Museum of Scotland.

## Quail-chicken Chimera

In the mid 1970s Nicole Marthe Le Douarin invented an embryo manipulation technology to produce chicken and quail chimera embryos. Her studies of quail-chicken chimeras have led to insights about higher animal nervous and immune systems. Her papers include "Tracing of Cells of the Avian Thymus through Embryonic Life in Interspecific Chimaeras" (1975). The photo illustrates a chimera born in the context of research carried out under the auspices of CNRS.

## Reconstructed Aurochs

Aurochs are depicted in many Paleolithic European cave paintings such as those found at Lascaux and Livernon in France. Heck Cattle, also called reconstructed aurochs, were developed in the early 20th century by the Heck brothers in Germany in an attempt to breed back modern cattle to their presumed ancestral form, the Aurochs. Modern cattle have become much smaller than their wild forebears: the height of a domesticated cow is about 1.4 meters, whereas aurochs could reach about 1.75 meters. Aurochs also had several features rarely seen in modern cattle, such as lyre-shaped horns set at a forward angle. Heinz Heck working at the Hellabrunn Zoological Gardens in Munich began creating the Heck breed about 1920. Lutz Heck, director of the Berlin Zoological Gardens, began extensive breeding programs supported by the Nazis during World War II to bring back the Aurochs. The reconstructed aurochs were useful to the Nazi propaganda machine, helping produce an idyllic Aryan history.

## Butterflies with modified wing patterns

In 2000 artist Marta de Menezes presented her work "Nature?," which consisted of live butterflies with modified wing patterns. During the pupation stage of the butterfly's life cycle, the artist interferes within the normal development of the wing. By cauterizing specific regions where the wing will originate, she changes the patterns of the wings of butterflies such as *Heliconius melphomene*, a brightly colored Latin-American species (photo). It is also possible to graft portions of tissues to other position in the same wing precursor or even into another butterfly's wings. There are no nerves in the wing, therefore the procedures do not cause pain. The modified butterflies have a normal life span and mating behavior. The artist states: "I do not have the intention of enhancing in any way Nature's design. Nor do I intend to make something already beautiful even more beautiful. I simply aim to explore the possibilities and constraints of the biological system, creating (inasmuch as it is possible) different patterns that are not the result of an evolutionary process."

## Sphynx Cat

The now apparently extinct Mexican Hairless (also called the New Mexican Hairless) is the precursor to the Sphynx Cat. Mexican Hairless written accounts go back to the early nineteenth century, but the first hairless "breed" properly recorded in the United States was a pair of Mexican Hairless given by Pueblo Indians in 1902 to a couple from New Mexico. It was claimed that these were the last survivors of an ancient Aztec breed of cat. The male, not yet sexually mature, was killed by dogs and the owners searched in vain for a hairless mate for the female. Throughout the twentieth century there were several reports of hairless or nude cats, but the modern Sphynx (Canadian Hairless) derives from Canadian cats of either the 1960s or 1970s. The history of the Canadian Sphynx is not continuous as the original bloodline has been lost. The hairless cat was bred in the Netherlands and elsewhere, continuously since the 1980s, but as the last authenticated Canadian Hairless of the earlier strain was lost the Sphynx breed was developed using Devon Rex. Devon Rex sometimes occurs with sparse fur. Rather than being totally hairless, the modern Sphynx derived from Canadian cats and other genetically compatible spontaneously occurring hairless cats has a light peach-fuzz on the skin and sometimes fur on the tail-tip.

## Longhorn clone

The Texas longhorn cattle is a Texas icon, bred more for show than for beef. In 2002, Starlight was the first Texas Longhorn in the world to be cloned.  Her tip-to-tip horn span measures 79 3/4 inches. A former world record holder for horn width in females, in 2006 Starlight was the number two female of the breed in horn span. She is a foundation matron for Zech Dameron's Clear Creek Pecan Plantation Ranch. It cost Dameron $31,000 to get five clones of Starlight. She cost $24,000. One of the clones didn't survive. In November 2002 four Starlight clones sold at a public auction.

## Hybrid Irises

Artist George Gessert's hybrid irises could not have occurred in the wild because they are derived from more than one species. He explains: "With animals, species cannot cross, or if they do, the hybrids that result are sterile, like mules, which are crosses between horses and donkeys. The same is not true with plants. With them the barriers between species may be genetic, but often are not genetic - e.g., geographical separation, or different bloom times. Nongenetic barriers between plant species are rarely or never breeched in the wild, but in the garden or greenhouse may disappear, or be overcome by manipulating the environment. Often the hybrids that result are fertile. Sometimes these fertile interspecific hybrids can be crossed with still other species and produce yet other kinds of fertile offspring". Gessert started to breed flowers in the late 1970s.

## Transgenic monkey

In 2001 scientists Anthony Chan, Gerald Schatten and colleagues created the world's first transgenic primate, a Rhesus monkey modified to incorporate the green fluorescent protein gene. Called Andi (anagrammatic acronym of 'inserted DNA'), the monkey carries but does not manifest the gene (that is, Andi does not glow). As with other in vitro fertilization (IVF) procedures in nonhuman primates, this one was relatively inefficient. Half of the fertilized eggs developed into embryos, and five pregnancies resulted from 20 embryo transfers, including one set of twins, which were miscarried (which is not unusual in monkeys with twins). Of the five pregnancies, three healthy monkeys were born, but only Andi had the GFP gene. One might conclude that, from a purely technical point of view, the project might have been easier to achieve in humans, for whom IVF technology is much more advanced.

## White Tigers

Contrary to popular belief, white tigers are not albino. Because of the rarity of the white tiger in the wild, white tigers can only exist in captivity by extensive inbreeding, which often leads to malformations, early deaths, still births and reduced intelligence (which helps explain why they are often used for entertainment). Although there are references to white tigers in the beginning of the nineteenth century, it was not until the 1950s that regular inbreeding began around the world. Due to the high market value for white tigers, breeders still inbreed white tigers today to ensure that offspring also exhibit the recessive gene.

## Cactus sculptures

Artist Philippe Obliger grafts cacti in whimsical or geometric forms to produce living botanical sculptures. Once the artist creates the grafts, the cactus tissue may take up to one year to bind with enough strength to support the form. Once stable, the grafted cactus behaves regularly and can live for many years. Featured here are the sculptures "Désespoir des singes" (Monkey desperation) [the ladder], and "Cube".

## Savannah cat

In 1986, breeder Judee Frank managed to crossbreed an African serval and a domestic cat, producing the first savannah cat (named Savannah). The Serval is a medium-sized African wild cat whose main habitat is the savanna. The length is 85 cm (34 in), plus 40 cm (16 in) of tail, which results, through crossing, in one of the larger breeds of domesticated cats. The Savannah cat's behavior is quite unique. For example: They may either chirp like their serval father, meow like their domestic mother, or do both, sometimes producing sounds which are a mixture of the two. Savannahs may also "hiss"—a serval-like hiss is quite different from a domestic cat's hiss, sounding more like a very loud snake hiss, and can be surprising to humans not familiar with a sound of this kind coming from a cat. The photo shows a fourth-month old F1 (first generation) Savannah cat.

## Antennapedia flies

Antennapedia flies are those in which a pair of legs replaces the antenna. In their 1998 paper, entitled "Control of antennal versus leg development in Drosophila," scientists Fernando Casares and Richard S. Mann, demonstrated that in the fruit fly antennae and legs are homologous structures that differ from each other as a result of a specific gene, which promotes leg identities by repressing unknown antennal-determining genes. To do so, they created Antennapedia (antenna-foot) mutant flies by transforming the antenna into leg and triggering antennal development elsewhere in the fly. The photos show one wild-type head standard phenotype (antennae) and two with the mutant phenotype (legs growing out of their head). Countless different mutant flies are regularly produced in labs worldwide in the course of genetics research.

## Blue rose

In 2004, the companies Suntory (Japan) and Florigene (Australia) announced the creation a blue rose. "Blue Roses" have long been a synonym for the impossible, but through genetic modification the companies succeeded in producing blue pigment on the petals of the flowers. Roses have been grown for more than 5,000 years. Varieties developed to more than 25,000 species and a wide number of colors exist including red, white, pink and yellow. For a long time, breeders have been trying to develop blue roses, but in rose petals genes encoding the enzyme that is necessary to create the blue pigment, "Delphinidin", are not functional. The transgenic roses created by Suntory and Florigene actually synthesize almost 100% Delphinidin (blue pigment) in their petals.

## Mouse with Human Ear

In 1997, scientist Joseph P. Vacanti put a mold resembling the shape of a human ear (made of biodegradable plastic) into the back of a mouse and seeded it with living cells. The mouse provided the nourishment for the growing cells, that took on the shape of the mold. Eventually the plastic mold dissolved, and the living cells generated new tissue in the shape of a three-dimensional ear. The prosthesis was not functional, that is, neither the mouse nor anyone else could use the ear to listen to anything. The purpose of the procedure was to contribute to the development of tissue-engineering, with the aim of regrowing ears and noses for humans.

## Cosmopolitan Chicken

The Cosmopolitan Chicken is a project by artist Koen Vanmechelen that aims to cross-breed chicken of different national races. In *The Variation of Animals and Plants under Domestication*, Darwin concluded that "all our breeds are probably the descendants of the Malayan or Indian variety of G. bankiva." Vanmechelen has studied the bankiva hen in Nepal and has hybridized chicken of many nationalities. The photos show: Mechelse Bresse (cross between the Mechelse Koekoek from Belgium and the Poulet de Bresse from France); Mechelse Giant (cross between Mechelse Redcap (third generation) x

Jersey Giant from the United States. Vanmechelen notes that "most of the chicken of the project (the bastards) live longer and became stronger after several generations."

## Zeedonk

The zeedonk is a cross between a zebra and a donkey. Shadow (photo) was bred at Colchester Zoo. He shares a large enclosure with, among other African herbivores, several Damara zebras. While the zebras form family groups and approach the perimeter fence, the zeedonk generally remains solitary and at a distance. Darwin recorded hybrids of the kind in the 19th Century. Zeedonks are usually bred by zoos and parks, mostly as attractions.

## Liger

A liger is a hybrid where the male is a lion and the female is a tiger. If the situation is reversed where the female is a lion and the male is a tiger, the offspring would be called a tigon. The liger is the world largest feline and can weigh over 900 pounds (408 kilograms) and stand almost 12 feet (3.6 meters) tall. The liger has both stripes and spots. The stripes are inherited from its tiger parent and the spots from the lion parent. Like tigers, but unlike lions, ligers enjoy swimming. There are records of ligers in the nineteenth century and ligers have been bred throughout the twentieth century. Male ligers and tigons are sterile but female ligers and tigons are fertile, and they can reproduce. There are known cases in which a liger was mated with a lion giving birth to cubs raised to adulthood. The liger Hercules (photo), lives in Jungle Island, an entertainment facility comprised of bird sanctuary, wildlife habitat, and botanical garden in Miami.

## Synthetic virus

In 2002 scientists at the State University of New York at Stony Brook produced from scratch a "synthetic" poliovirus, known as responsible for the disease poliomyelitis. The research was financed by the Pentagon as part of a program to develop bio warfare countermeasures. Dr. Eckard Wimmer, the lead scientist in the study, said they made the virus to send a warning that terrorists might be able to make biological weapons without obtaining a natural virus. Because of its short genome and its simple composition poliovirus is regarded as the simplest significant virus.

## Semi-Living Worry Dolls

The Tissue Culture & Art Project (Oron Catts, Ionat Zurr and their associates) created in 2000 an art installation entitled "Tissue Culture & Art(ificial) Womb." In this work the artists grew their own versions of the Guatemalan Worry Dolls in a bioreactor. Traditionally, Native-Guatemalans teach their children to tell their worries to one of these small dolls at bedtime so that the doll will solve their worries overnight. The artists produced seven semi-living worry dolls with degradable polymers and surgical sutures. They explain: "The dolls were sterilized and seeded with endothelial, muscle, and osteoblasts cells (skin, muscle and bone tissue) that are grown over/into the polymers. The polymers degrade as the tissue grows. As a result the dolls become partially alive. The process, in which the natural (tissue) takes over the constructed (polymers), is not a "precise" one. New shapes and forms are created in each instance, depending on many variants such as the type of cells, the rhythm of the polymer degradation and the environment inside the artificial womb (bioreactor). It means that each doll transformation cannot be fully predicted and it is unique to itself."

## Quagga

The quagga is an extinct subspecies of the plains zebra, which was once found in great numbers in South Africa's Cape Province and the southern part of the Orange Free State. It was distinguished from other zebras by having the usual vivid marks on the front part of the body only. In the mid-section, the stripes faded and the dark, inter-stripe spaces became wider, and the hindquarters were a plain brown. The name comes from a Khoikhoi word for zebra and is onomatopoeic, being said to resemble the quagga's call. After the very close relationship between the quagga and surviving zebras was discovered, the Quagga Project was started by Reinhold Rau in South Africa to recreate the quagga by selective breeding from plains zebra stock, with the eventual aim of reintroducing them to the wild. This type of breeding is also called breeding back. On 20 January 2005 the most quagga-like foal, called Henry, was born (photo, with his mother). The striped area of its body is not only much reduced, but the body stripes themselves are considerably narrower and fainter than usual, more so than in some of the museum specimens of the former quagga population. However, there are some stripe remnants on the hocks. Such remnants are not present on the museum specimens. In early 2006, it was reported that the third and fourth generations of the project have produced animals which look very much like the depictions and preserved specimens of the quagga, though whether looks alone are enough to declare that this project has produced a true "re-creation" of the original quagga is controversial. The Quagga Project is an attempt by a group of dedicated

people in South Africa to bring back the Quagga
(Equus quagga quagga) from extinction and
reintroduce it into reserves in its former habitat.

## Hypoallergenic cat

In 2006 the world's first specially bred hypoallergenic
cats went on sale in the US at the initial price of
$3,950 USD. A tiny protein particle, the "Fel d 1"
allergen is found mainly in the cat skin flakes and
saliva. The protein is produced in the cat salivary
glands and sebaceous glands of the skin. Cats are
fastidious groomers, so they deposit the Fel d 1
protein on their fur by licking themselves. The
allergen is so small it can remain airborne for
months. Using "gene silencing" technology, Allerca is
able to suppress the production of the Fel d1
allergen. The company says the animals will not
cause the red eyes, sneezing and even asthma
triggered by cat allergy, except in the most acute
cases. Allerca cats are medium sized cats averaging
between ten and fifteen pounds (4 to 8 kg). They are
fully mature at approximately three years of age and
have a long life expectancy.

## Catalina Macaws

The Catalina Macaw is a hybrid produced by
breeding a Scarlet macaw and a Blue and Gold
macaw. Generally, the female of the pair is the Blue
and Gold, since Scarlet macaw females are more
difficult to locate. The Catalina macaw usually has a
yellow-orange chest with green on the top of the
head and green to green-blue, with shadings of
orange, on the back of the neck and back. The
Catalina Macaws have a long tapered tail similar to
that of the Scarlet Macaw. When breeding two

hybrid macaws the results can vary in the offspring.
The personality characteristics of the Catalina
Macaws are a meshing of the Blue and Gold Macaw
and the Scarlet Macaw. The friendly disposition of
the Blue and Gold tempers the more high-strung
tendencies of the Scarlet. They tend to be very
intelligent, playful, and excellent talkers. They have a
natural curiosity, a willingness to learn, and often
become very bonded with members of their family.
As with any of the macaws, their handling as babies
plays an important role in the development of their
social skills and personality.

## Laufberger's Axolotl

The Axolotl (Ambystoma mexicanum) is an aquatic
salamander native to Mexico. The name axolotl
comes from Nahuatl; in Spanish it is called ajolote. It
is noted for its appearance and its neoteny, meaning
that it remains in its aquatic larval form even as a
sexually mature adult, and does not undergo
metamorphosis into a terrestrial form. In 1913 Vilém
Laufberger of Germany published his discovery: He
used thyroid hormone injections to induce an axolotl
to grow into a terrestrial adult salamander. Under the
influence of the artificially injected hormone, the
axolotl grew up and an adult salamander was
created, quite possibly for the first time. The
experiment was repeated by the Englishman Julian
Huxley, who was unaware it had already been done,
using ground thyroid hormones. Since then,
experiments have been done often with injections of
iodine or various thyroid hormones used to induce
metamorphosis. Artificial metamorphosis
dramatically shortens the axolotl's lifespan, if they
survive the process. A neotenic axolotl will live an

average of 10–15 years (though an individual in Paris is credited with achieving 25 years), while a metamorphosed specimen will scarcely live past the age of five. The adult form resembles a terrestrial Mexican Tiger Salamander, but has several differences, such as longer toes, which support its status as a separate species. The photos show the axolotl during various stages of development.

## Schwarzenegger Cow

The so-called 'Schwarzenegger cow' is a variant of the Belgian blue cattle breed, with double the muscle of an average cow. This is due to a mutant gene that suppresses the production of Myostatin, a protein that normally inhibits muscle growth after a certain point of development. Pure Belgian Blue carry two copies of the gene and are thus known for their "double muscling" phenotype (hence the Schwarzenegger epithet). The 'Schwarzenegger cow' strains have been produced through breeding.

## Mushroom Sculptures

Artist Philip Ross grows mushroom sculptures by manipulating a type of fungus called Ganoderma lucidum (also known as Reishi or Ling Chi). Mushrooms "bloom" much like flowers do at certain times of the year. Hidden underground (and thus invisible) is the root-like network of filaments (hyphae) making up the non-reproductive part of the body of a fungus. This underground component of the fungus is called mycelia. The mushroom is the reproductive organ of the fungus. Ross cultures Ganoderma lucidum and casts the fungus in the shapes he creates. Ross explains: "After a month the solidified form can be slipped out of its chicken wire container and maintain its shape. The fruit of the organism (the thing that looks like an antler or shelf) is coaxed to grow into different shapes by changing its relationship to light, gravity and oxygen/co2 levels. The entire process takes from nine months to a year to complete.".

## Bizzarria Orange

The bizzarria orange is a chimera, that is, a life form with cells from two different beings; in this case from bitter orange and the citron. The bizzarria is a graft-hybrid and presents the remarkable spectacle of these two different fruits blended into one. Historically, it was first recorded in 1674 by Pietro Nati, the director of the Botanical Garden of Pisa. He wrote that it was a gardener at the villa Torre degli Agli, belonging to Marquis Panciatichi, outside of Florence, that created it. It was Nati who gave it the name Bizzarria. In the seventeenth century, the bizzarria fruit was drawn by Florentine painter Baldassare Franceschini, called Volterrano. Darwin wrote in the book *The Variation of Animals and Plants under Domestication:* "The gardener who in 1644 in Florence raised this tree, declared that it was a seedling which had been grafted; and after the graft had perished, the stock sprouted and produced the bizzarria. Gallesio, who carefully examined several living specimens and compared them with the description given by the original describer, P. Nato (11/100. Gallesio 'Gli Agrumi dei Giard. Bot. Agrar. di. Firenze' 1839 page 11. In his 'Traite du Citrus' 1811 page 146, he speaks as if the compound fruit consisted in part of a lemon, but this apparently was a mistake.), states that the tree produces at the same time leaves, flowers, and fruit identical with the bitter orange and with the citron of Florence, and likewise compound fruit, with the two kinds either blended together, both externally and internally, or segregated in various ways. This tree can be propagated by cuttings, and retains its diversified character."

## Methuselah Worm

In 2003 scientist Cynthia Kenyon of the University of California at San Francisco and her colleagues created transgenic worms (Caenorhabditis elegans, or simply C. elegans) that lived six times longer than normal worms and remained active for most of their lives. The Methuselah Worm can live 144 days—the equivalent of a human reaching his 500th birthday. To create the long-life worms the scientists changed the genes in C. elegans that affect the activity of insulin and removed gonad tissue, which affects endocrine hormone levels. "These life-span extensions, which are the longest mean life-span extensions ever produced in any organism, are particularly intriguing," the team explained, "because the insulin/IGF-1 pathway controls longevity in many species, including mammals."

## Alba, the bunny

In 2000 Eduardo Kac created the fluorescent rabbit called Alba. The artist inserted in a rabbit fertilized ovum the fluorescence genes found in the jellyfish Aequorea victoria (aka crystal jelly), native to the west coast of North America. The fluorescence feature was transferred to Alba. As a result, when exposed to blue light of a specific frequency, Alba glows green. Traditionally, creatures made of the parts of multiple animals were firmly in the realm of mythology. However, with his work "GFP Bunny," in the context of which Alba was born, Kac transformed myth into medium. She is a transgenic mammal. The artist intended to pass from legend to life, thus bringing into the world a new being of his own creation. For Kac, bio art allows him to create not objects, but real subjects. "GFP Bunny" was censored by the director of the lab where she was born, in Jouy-en-Josas, France. Contrary to what was originally agreed (i.e., Kac's stated goal of bringing Alba home), she was never allowed to leave the lab.

## Avital Ronell

Avital Ronell is Jacques Derrida Chair at The European Graduate School, Saas-Fee, Switzerland, and professor and chair, German Department, New York University. Ph.D., (Princeton University). Former performance artist and professor of comparative literature, University of California at Berkeley. Avital Ronell bridges the European and American theory traditions and has contributed to the deconstructive reading of technology and communication as well as ethics and aesthetics. Author of *Dictations; The Telephone Book; Crack Wars; Finitude's Score*. Ronell studied at the Hermeneutics Institute in Berlin with Jacob Taubes, ultimately earned her doctorate at Princeton University, and then worked with Jacques Derrida and Hélène Cixous in Paris. From her first book, *Dictations*, through her latest one, *Stupidity*, Ronell calls the established questions into question, zooming in on whatever "withdraws from immediate promises of transparency or meaning" and/or tracking what she calls the "rhetorical unconscious of a text". A hybrid of high theory and street talk, Ronell's texts are remarkable both for what they say and for the extraordinary way in which they say it. Each of her works goes after a seemingly recognizable and knowable signifier (Goethe, the telephone, the drug addict, the television, the test, the greeting, stupidity, etc.) but then tracks it so closely that it quickly becomes unrecognizable, exceeding its object-status, overflowing itself as a concept. Explicitly breaking with scholarly tradition, a tradition that values mastery and certitude, Ronell engages her "object" of study at the level of its finitude, of its radical singularity. Whatever the topic at hand, Ronell's overarching concern is with an "ethics of decision" for this postfoundational era—an era in which all the transcendental navigation systems are down: "To the extent that one may no longer be simply guided—by Truth, by light or logos—decisions have to be made." It's only in certitude's interruption that meaning's inappropriability is exposed; and it's only in that exposure that an ethics of decision becomes available. (Excerpted from a text by Diane Davis)

## Eduardo Kac

Considered one of today's leading artists, Kac opened at the dawn of the twenty-first century a new direction for contemporary art with his "transgenic art"—first with a groundbreaking net installation entitled Genesis (1999), which included an "artist's gene" he invented, and then with his fluorescent rabbit called Alba (2000). From his first online works in 1985 to his current convergence of the digital and the biological, Kac has always investigated the poetic, philosophical and political dimensions of communication processes. Equally concerned with the aesthetic and the social aspects of verbal and non-verbal interaction, in his work Kac examines linguistic systems, dialogic exchanges, and interspecies communication. Kac's work has been exhibited internationally at venues such as Exit Art and Ronald Feldman Fine Arts, New York; Maison Européenne de la Photographie, Paris, and Lieu Unique, Nantes, France; OK Contemporary Art Center, Linz, Austria; Seoul Museum of Art, Korea; and Museum of Modern Art, Rio de Janeiro. Kac's work has been showcased in biennials such as Yokohama Triennial, Japan, Bienal de São Paulo, Brazil, and Gwangju Biennale, Korea. His work is part of the permanent collection of the Museum of Modern Art in New York, the ZKM Museum, Karlsruhe, Germany, and the Museum of Modern Art in Rio de Janeiro, among others. His public art commission from the University of Minnesota is in the permanent collection of the Weisman Art Museum, Minneapolis. His writings are collected in three volumes: *Light & Letter. Essays in art, literature and communication* (Rio de Janeiro: Contra Capa, 2004); *Telepresence and Bio Art: Networking Humans, Rabbits and Robots* (Ann Arbor: University of Michigan Press, 2005) and *Hodibis Potax* (Ivry-sur-Seine, France: Édition Action Poétique/ Maribor, Slovenia: Kibla, 2007). His anthology *Signs of Life : Bio Art and Beyond* (Cambridge, MA: MIT Press, 2007), documents the history and theory of bio art. His work is documented on the Web in eight languages: http://www.ekac.org.

*cover images*
George Gessert *Hybrid Iris (cover)*
Philip Ross *Mushroom sculptures (back)*


*design*
Lili Fleury
*layout*
Eleinad Ereivir


THIS SERIES EDITED BY DANIELE RIVIERE

*in the same series :*
"School Spirit" by Pierre Huyghe & Douglas Coupland
"Spread Wide" by Kathy Acker, Paul Buck, John Cussans, Rebecca Stephens


© DISVOIR, 2007
1Cite Riverin
F-75010 PARIS
http://www.disvoir.com


ISBN: 9782914563345


PRINTED IN EUROPE

printed by
POLICROM
Barcelona
Spain
october 2007